Myung Ja Kim is currently a postdoc teaching fellow in Northeast Asian Politi African Studies (SOAS), University of Lc at the Politics Department at SOAS whe Scholarship Award. Her MA in Internatior ... ..ompieieu at the School of International Service, American University in Washington, DC. She has been a guest lecturer in Korean Studies at Tübingen University, and her work has been published in the *Journal of Asian Security and International Affairs*. She founded and was president of the NGO World Tonpo Network, Tokyo, an organization that seeks the peaceful unification of North and South Korea.

'This is a wonderful book about the Zainichi Koreans – a group that was identified neither as Japanese nor as foreigners. There is rich history in these pages but also useful arguments and lessons about how host nation-states treat diaspora groups.'

Victor Cha,
D.S. Song-KF Professor of International Affairs and Government, Georgetown University and author of *Alignment Despite Antagonism: The US–Korea–Japan Security Triangle*

'This book will ignite healthy and timely academic debate and advance our knowledge.'

Seung-Young Kim,
Senior Lecturer,
School of East Asian Studies, University of Sheffield

The Korean Diaspora in Postwar Japan
Geopolitics, Identity and Nation-Building

MYUNG JA KIM

I.B. TAURIS
LONDON • NEW YORK • OXFORD • NEW DELHI • SYDNEY

I.B. TAURIS
Bloomsbury Publishing Plc
50 Bedford Square, London, WC1B 3DP, UK
1385 Broadway, New York, NY 10018, USA

BLOOMSBURY, I.B. TAURIS and the I.B. Tauris logo
are trademarks of Bloomsbury Publishing Plc

First published in Great Britain 2017
Paperback edition published 2020

Copyright © Myung Ja Kim, 2017

Myung Ja Kim has asserted their right under the Copyright,
Designs and Patents Act, 1988, to be identified as Author of this work.

All rights reserved. No part of this publication may be reproduced or
transmitted in any form or by any means, electronic or mechanical,
including photocopying, recording, or any information storage or retrieval
system, without prior permission in writing from the publishers.

Bloomsbury Publishing Plc does not have any control over, or responsibility for,
any third-party websites referred to or in this book. All internet addresses given
in this book were correct at the time of going to press. The author and publisher
regret any inconvenience caused if addresses have changed or sites have
ceased to exist, but can accept no responsibility for any such changes.

A catalogue record for this book is available from the British Library.

A catalog record of this book is available from the Library of Congress.

ISBN: HB: 978-1-7845-3767-8
PB: 978-0-7556-0103-5
ePDF: 978-1-7867-3185-2
eBook: 978-1-7867-2185-3

International Library of Twentieth Century History 94

To find out more about our authors and books visit
www.bloomsbury.com and sign up for our newsletters.

Contents

List of Illustrations — vii
Acknowledgements — viii
List of Abbreviations — ix

Introduction — 1
The Puzzle: Contradictory Identities within the Zainichi Diaspora — 1
Korean Migration in Japan: Becoming Zainichi — 10
The Relationship between Diaspora and Nation State — 14
Primordialist Perspectives on Nation — 17
Constructivist Views on Nation — 21
Modernization Vision on Nation-Building — 24
Geopolitics, Making Nation-Building Policies and the Zainichi Diaspora — 25

1 Alliance Cohesion, Diaspora and Nation-Building Policies — 30
Nation-Building Policies via Interstate Relations — 31
Diaspora and Divided External Powers — 33
State Capacity and Variations of Power Configurations under Asymmetric Alliances — 36
Diaspora as an Alternative to Nation State — 39
External involvement vs Domestic Politics — 41
Possible Causal Pathways — 43
Research Design — 46

2 The Zainichi Diaspora: From the Shadow of Japan's Colonial Legacy — 51
Introduction — 51
The Annexation of Korea — 54
Korea's Response to Japanese Colonization Policies — 59

CONTENTS

Imperialism under the Slogan 'Japan and Korean as One Body'   63
The Legacy of Japanese Colonization   68

3 No Alliance and a Strong Historical Legacy: Exclusionary Policies towards the Zainichi in the Post-World War II Era (1945–64)   73
Introduction   73
The Korean War and the US–ROK Alliance   79
The San Francisco Peace Treaty and the US–Japan Alliance   89
Exclusionary Policy toward the Zainichi in the Post-World War II Era (1945–64)   98
Conclusion   112

4 Alliance Cohesion Matters: Japan's Policy towards the Zainichi during the Cold War Era (1965–80s)   115
Introduction   115
Empirics: Alliance Cohesion as a Causal Factor in Japan's Policy towards the Zainichi Diaspora in the Cold War Era (1965–80s)   118
Conclusion   160

5 Does Alliance Cohesion Still Matter in the New Post-Cold War (1990–2014)?   162
Introduction   162
The Post-Cold War Era (1990–2014)   166
Conclusion   202

Conclusion   209
Geopolitics and Shifting Policies towards the Zainichi Diaspora   209
Factors Identified by the Zainichi Diaspora Case Study   211
Accommodation Policies towards Divided Diaspora Groups and Variation in Outcomes   214
Policy Implications   217

Notes   224
Bibliography   254
Index   271

List of Illustrations

Figures

1.1	Playing the diaspora card – the role of interstate alliance in the treatment of diaspora groups	40
1.2	Causal pathways: alliance cohesion, interstate relations and policy toward the Zainichi	44
2.1	Population of ethnic Koreans in Japan, according to the census, 1910–45	52
4.1	Predicted policy-shifts toward the Zainichi during the Cold War era	118
4.2	*Gaikokujin Tokei* (statistics on foreigners in Japan)	157
5.1	North Korean missile ranges	189

Maps

3.1	The 38th parallel	75
3.2	Korea's geographical position in Asia	77

Acknowledgements

This book is, to a large extent, driven by my own search for identity – something that has interested me for as long as I can remember, and which has led me to think deeply about who I am now and how that may shape my future. The ambiguous identity of post-colonial Koreans has held a fascination for me since my childhood days in Japan. In the course of my journey I have been privileged to study at SOAS, where I was influenced by passionate scholars such as my supervisor, Dr Fiona B. Adamson. Without her assistance, the completion of this book would not have been possible. I have learned about diaspora politics in particular detail from Dr Adamson; from the frequent interactions with her during my PhD course I hope, one day, to obtain answers to my own questions of identity which I can then use as a foundation for further academic inquiry. The ontological mechanisms discovered and discussed in this book (which are still open to further exploration) would be applicable to many other regions in the world, and my ambitions are to pursue an academic career in which I will be offered the chance to explore these possibilities to the best of my abilities.

I would like to acknowledge the support of the Meiji-Jingu Intercultural Research Institute in Tokyo. I was able to fully concentrate on my studies throughout my thesis thanks to the support of this institute via my scholarship.

I would also like to thank Dr Tat Yan Kong and Dr Harris Mylonas, whose advice and comments have greatly enriched my work.

Finally, I would like thank my family, especially my son, Yuki, who gives me the inspiration to continue this work. I would also like to thank my parents, who lived in a host country and raised their children with determination and incredible effort.

List of Abbreviations

ABM	Anti-Ballistic Missile
ARO	Alien Registration Ordinance
CIS	Commonwealth of Independent States
DPRK	Democratic People's Republic of Korea (North Korea)
EEZ	Exclusive Economic Zone
FEC	Far East Commission
GDP	Gross Domestic Product
GHQ	General Headquarters
GNP	Gross National Product
GS	Government Section
GSDF	Japan Ground Self-Defence Force
HNS	Host Nation Support
ICRC	International Committee of the Red Cross
JCS	Joint Chiefs of Staff
JDA	Japanese Defence Agency
JSDF	Japan Self Defence Forces
JSP	Social Democratic Party of Japan
JUSMAG-K	US Military Assistance Group-Korea
KCIA	Korean Central Intelligence Agency
KDP	Korean Democratic Party
KEDO	Korea Energy Development Organization
KMAG	Korean Military Advisory Group
KPR	Korean People's Republic
LAZAK	Lawyers Association of Zainichi Koreans
LDP	Liberal Democratic Party
LWR	Light Water Reactor

ABBREVIATIONS

NPT	Non-Proliferation Treaty
ODA	Official Development Assistance
OECD	Organisation for Economic Co-operation and Development
PRC	People's Republic of China
RCC	Resolution and Collection Corporation
ROC	Republic of China (Taiwan)
ROK	Republic of Korea (South Korea)
SALT	Strategic Arms Limitation Talks
SAR	Search-and-Rescue
SCAP	Supreme Commander of the Allied Powers
SDF	Self-Defence Force
SDI	Strategic Defence Initiative
UNCLOS	United Nations Convention on the Law of the Sea
UNTCOK	United Nations Temporary Commission on Korea
USMGK	US Military Government in Korea
USSR	Soviet Union
WMD	Weapons of Mass Destruction

Introduction

The Puzzle: Contradictory Identities within the Zainichi Diaspora

After the end of World War II, the Japanese elite began to prioritize national integration as part of their nation-building policy. But the ambiguous ethnic identification of post-colonial Koreans who were living in Japan at the time is proof that Japan's policy developed in a form that was neither linear nor coherent. Japanese citizenship for members of the Zainichi diaspora was not made impossible but neither was it made easy, as Zainichi Koreans still possessed legal claims to their Korean homeland despite having become culturally Japanese. As John Lie argues, 'non-Japanese Japanese' had not been granted a place in Japanese society as far as state bureaucrats were concerned – either in the present or in the past.[1] In other words, a diaspora of Zainichi Koreans should not exist in Japanese society due to the monoethnic nation building policy adopted by Modern Japan in the postwar period. The Zainichi are of particular interest as a case study, because unlike ethnic Koreans in China and Russia as well as Korean Americans they represent the only Korean migrant group that has not been granted citizenship by its host state. Having been brought here principally as a source of cheap wartime labour either voluntarily or forcefully, the Japanese expected the group to return to Korea en masse, but the geopolitical complexities that existed on the Korean peninsula during the postwar years meant that the Zainichi were unwilling to return – or found themselves unable to – and stayed in Japan in a state of limbo, remaining outside the protection of Japanese citizenship in a country that wasn't sure what to do with them. Most Zainichi Koreans came originally from the southern provinces of Korea and retained their Korean nationality, but now they speak Japanese rather than Korean and their appearance, culture and fashion represent Japanese styles. They even have Japanese names; in other words, they have become highly acculturated to Japanese society,[2] but despite the fact that the naturalization rate has increased since the 1990s, there are still around 400,000 Zainichi in Japan who maintain their Korean nationality but have no intention of moving back to Korea.[3]

The contemporary Zainichi are a far from homogeneous group; they reflect diverse perspectives, attachments and configurations of identity. Zainichi social movement activists have lobbied for the half-measure of acquiring voting rights in local assemblies instead of trying to acquire fully fledged national citizenship. As a counter movement to offering foreigners voting rights in Japanese elections, some conservative politicians in 2001 participated in the preparation of a bill to grant special permanent residents (principally Zainichi Koreans) Japanese nationality,[4] but this quiet voice was soon stilled. Historically, all policies that might lead to assimilation were opposed by the two ethnic Korean associations, *Mindan* (pro-South) and *Chongryun* (pro-North). But now *Mindan* supports local voting rights for permanent residents, while members of *Chongryun* oppose suffrage that might weaken any political ties they have to North Korea.[5] Despite the influence of the two associations, most gains in civil and social rights for Koreans have come about through the influence of smaller, politically independent civic groups.[6] Zainichi often take on Japanese name-forms or use Japanese aliases, allowing them to blend in with the Japanese in everyday social encounters,[7] but they soon realize that their ambiguous identity masks an underlying disadvantage in terms of their own social lives. To perpetuate such a contradictory ideology, clinging to legal claims on South or North Korea with no intention of returning to a homeland that has been radically divided since their departure brings problems when it comes to making critical choices in their lives such as marriage or finding employment.

From the end of World War II to the present, Japan's policies towards foreign settlers have changed from exclusion to inclusion and back again, but institutional barriers still remain. What's more, the process of integration has not been conducted evenly across the diaspora, which implies some political manoeuvring between the Japanese state, the Zainichi factions and the two Korean regimes. From the Japanese elite's nation-building perspectives, however, homogenization is an imperative part of the process, and must be achieved before their nation-building efforts can be considered complete. The political and social marginalization of immigrant populations can be seen as a potentially destabilizing factor which must be overcome by the liberal democracies of the host state.[8] But in this case, neither the Japanese elites nor the Zainichi themselves are fighting to change the rule that citizenship is a matter of choice – even for the long-term immigrants who date their lives in Japan back to the colonial era. Despite this, the Zainichi diaspora has no intention of repatriating to Korea. So what factors generate and perpetuate the contradictory identity of the Zainichi? What drives Japanese policy makers to allow the continuing existence of such an ambiguous diaspora group that clings to legal claims to Korea whilst maintaining a culturally Japanese identity?

The plight of Korean communities in modern Japan has been discussed in both political and academic terms (Mitchell 1967, Lee and De Vos 1981, Weiner 1989, Ryang 1997, Hicks 1997, Chapman 2008, Lie 2008, Chung 2010). The peculiarities of the Korean diaspora, however, have been explored only from the Japanese domestic viewpoint, and have not been examined beyond this scope, particularly in terms of international relations. But a closer examination reveals a number of unique and contradictory features regarding modern Japanese identities – as well as those of North and South Korea – both within the Zainichi diaspora and regarding attitudes towards it. For example, unlike other Korean diasporas in China, Russia and the US, the Zainichi remain split along the political lines that reflect its divided homeland. *Mindan* and *Chongryun*, the two divisive organizations mentioned above, have competed for the loyalty of Koreans in Japan since the late 1940s. Meanwhile Koreans in China overwhelmingly sided with the Communists at the outbreak of the civil war between the Chinese Communists and Chiang-Kai-shek's Nationalists. As far as allied relations between China and North Korea are concerned, China's ethnic Korean diaspora publicly praised North Korean leader Kim Il Sung as a great patriot and a fighter for independence.[9] In contrast, the majority of ethnic Koreans in the USSR (*Koryŏ Saram*) had no such political ties, and their connections to Korea have always been weak.[10] Thus in no other country with a large Korean minority has that diaspora been as sharply divided as it is in Japan because of the polarizing strength of the two rival Koreas' influences on the Zainichi diaspora.

The Korean diasporas in China and the USSR can provide us with important variables – such as regime type or political ideology – which could affect nation-building policies towards minority groups. For the *Josŏnjok* (the Korean diaspora in China) there was no clear alternative. According to Enze Han, Socialism and the Communist Party were the only possible alternatives, because the *Josŏnjok* were exposed only to China and North Korea at the time, both of which were allied communist countries.[11] Soviet Koreans (*Koryŏ Saram*) played a key role in the creation of the Democratic People's Republic of North Korea (DPRK), but the history of the Korean diaspora in the former USSR shows that political ideology was neither a necessary nor a sufficient condition for the host state's nation-building policy decisions. In the 1950s and 1960s, for example, Kim Il Sung attacked real and potential rivals for their alleged pro-Soviet or pro-Chinese sympathies. He used nationalism as a weapon, casting up their backgrounds in the USSR or the Chinese Communist Movement.[12] Many purged cadres returned to the Soviet Union and China with no political ties to North Korea.[13] It is self-evident that there was no connections with South Korea either, especially during the Cold War era. From the perspectives of China and the former USSR, political ideology matters in terms of nation-building policies towards the different ethnic residents at some point, but is not a sufficient explanatory factor in its own right. What seems more important to policy makers is the perceived degree of threat; unlike the Zainichi, the ethnic Korean diasporas in the former USSR and China have weak relations with their external motherlands and no ideological choices. As such, they provided no serious threat towards nation-building processes in China or the USSR.

If a regime or its political ideology remains an important variable, it is hard to see why the Zainichi diaspora – which came primarily from the southern part of the Korean peninsula and remained in Japan after August 1945 – were treated as third-country nationals. They were seen as neither Japanese nor foreigners,[14] and were clearly defined as a special category. Their South Korean homeland and their new Japanese home were both supported by the same liberal democracy, the US. Ironically, it was the US occupying authorities who played a critical role in terms of the treatment of the remaining Koreans in Japan, as it was they who re-classified the Zainichi from 'imperial subjects' to the special category of third-country nationals.[15] In a statement released on 19 November 1946, the SCAP changed its official view slightly by stating that Koreans in Japan were to be regarded as Japanese nationals until Korea's legitimate government recognized their Korean nationality. Since the Republic of Korea (South Korea) and the Democratic People's Republic of Korea (North Korea) were not established until 1948, so Koreans in Japan between 1945 and 1948 were considered stateless.[16] The SCAP tacitly approved this treatment for security reasons, maintaining that legal jurisdiction over Koreans in Japan should be exercised by the Japanese authorities. But the Japanese government began to treat Koreans as *de facto* 'foreigners' by manipulating the situation, restricting their rights in areas of political participation and social welfare.[17] In response, left-wing activists within Japan (who were perceived as security concerns by US and Japanese governments) played a prominent role in organizing Korean residents. The leading Korean-Japanese association – the *Choren*, established after World War II – had close ties to the Japanese Communist Party and was favourably inclined towards North Korea.[18] During the 1945–52 US occupation of Japan, the Korean population was perceived as a troublesome minority that might threaten the homogenization of the nation-building process that was moving towards a 'mono-ethnic' Japan which conformed to models of the capitalist-liberal democracies which were standing against communism at the time.

But the treatment of the Zainichi gives rise to a broader question: what subsequently drove the Japanese elites' perception of the Zainichi diaspora to change from a single category of threat into two opposed political categories depending on perceptions of allegiance, in which one side was supported by an allied state and the other by enemy state? How do we explain the variation in state policies towards the two different groups among the Zainichi diaspora? The argument that external and internal factors played significant roles in developing a competing Korean identity within the diaspora and in reflecting East Asian geopolitical power configurations at critical fixed points in history has been under-researched, although it holds an increasing relevance today.[19] Existing literature on the Zainichi focuses on their status as a minority within Japan's domestic political environment, and issues of citizenship only consider the Japanese domestic context.[20] This book differentiates the Zainichi from other host state minorities such as Burakmin, Ryukyuan/Okinawan and Ainu in terms of recognition insofar as the Zainichi maintain a 'homeland' – either in physical terms or as a psychological entity – as well as an imagined locus of emotional identification. By taking this approach, the trajectory of Zainichi Korean history can be analysed from a very different perspective. Of the various places of the world in which US military commitment has been extensive, the Korean peninsula has been one of the most controversial since the end of World War II.[21] In comparison with the strong influential role of US foreign policy in South and North Korea during the Cold War, Japanese foreign policy has often been characterized as passive in following American leadership.[22] However, as Seung-Young Kim argues, we have witnessed recurrent attempts by Japan to take independent initiatives over issues on the Korean peninsula and in East Asia as a whole.[23] These initiatives were driven by Japan's perception of the US-centred alliance system and its relations with the two Koreas, which was subject to constant changes in international politics in the Asia–Pacific region. In other words, from Japan's perspective, foreign diplomacy towards the two Koreas has not been always consistent with US policies. Given their ambiguous identities, the Korean diaspora in Japan has inevitably been involved in Japan's strategic diplomacy toward East Asian geopolitics from the end of war to the present. The importance and relevance of this book is thereby justified, as there is little extant literature exploring the role of the Zainichi in terms of international relations.

Recently, growing numbers of US and Japanese policy makers have begun to focus on the 'super-sizing' of North Korea in terms of its continuing military threat.[24] However, the transnational space in which the Zainichi diaspora functions has been ignored for a long time, and only came to a head when North Korea launched a three-staged Taepo-dong ballistic missile directly over Japanese airspace on 31 August 1998, an action which sent international shockwaves reverberating across the world, with Japan feeling particularly threatened. Current academic research has begun to examine the transnationalism that plays an important role in any threat to Japan's national security and regional peace by exploring the role of the pro-North Korean ethnic organization *Chungryun*, which became a powerful agency working within Japan to provide the North Korean regime with funds and intelligence.[25] As such, this book contributes to uncovering and exploring a transnational space in which the Zainichi's changing roles within Japan's borders have been facilitated by host state policy shifts that reflect the external security environment and the impact on that environment which these changing roles have brought. A diaspora can do what a quiet minority cannot; it can function in transnational space, thereby providing leverage that can be exploited by individuals or collective actors who make up both host state and homeland elites.

Recent diaspora scholarship substantiates the growing importance of diasporas in international politics, explaining that diasporas have emerged as new and potentially powerful actors, influencing political mobilization in the management, escalation and settlement of interstate conflicts.[26] Diasporas can also provide influential lobbying in their host state on behalf of their homeland[27] – in more extreme cases helping to create new transnational nations, as the activities of the Armenian, Tamil, and Croatian diasporas have historically shown.[28] But how does a particular ethnic group become engaged in ethnic mobilization across a transnational space, ending up as a group that involves itself in international relations and political struggles? Some argue that diasporic motivation stems from a stronger ethnic and national attachment to homeland.[29] Others say that it is caused by host state policies that have prevented seamless integration.[30] The question is closely related to diaspora literature that concentrates on the reasons diasporas maintain homeland ties, and needs to be approached by exploring nation-building literature. Japan has in fact pioneered two different concepts of nationhood[31] between the Meiji era and the end of World War II; one was a multi-ethnic empire, the other a homogeneous mono-ethnic nation state. The second type was accompanied by the disintegration of Japan's empire in the context of what Rogers Brubaker calls the 'unmixing of people.'[32] At the end of World War II, there were approximately 2 million Koreans, 200,000 people from the Ryukyu archipelago (Okinawans), 56,000

Chinese and 35,000 Taiwanese in Japan. Several post-imperial migrations ensued when the ruling ethnic majority of Japan's multi-national empire was abruptly transformed by a postwar reconfiguration of political authority along national lines due to the sudden shrinkage of its geopolitical space. In this context, Brubaker makes the important point that the reconfiguration of political space along national lines did not necessarily entail a corresponding redistribution in actual population.[33]

This book seeks to discover an answer to questions relating to the conditions under which host states treat an ambiguous diaspora which identifies its essential belonging as being divided, and whether that treatment is inclusive or exclusive. Harris Mylonas argues that interstate relations between the diaspora group's host country and homeland will affect the perceptions of the host state's ruling elites towards that diaspora.[34] Incentives for policy making towards the diaspora will therefore depend on whether host and home states belong to the same alliance.[35] However, by building on the logic of Mylonas' nation-building theory, this book will also consider the idea that host states will sometimes prioritize their national concerns by manifesting an interest in forging closer ties with enemy states or alliances of enemy blocs. In the US in 2015, for example, the Obama administration prioritized normalizing relations with its one-time enemies, Iran and Cuba, at the expense of fostering longstanding friendships with nations such as Saudi Arabia and Israeli.[36] In the same way, Japanese Prime Minister Tanaka (1972–4) also accepted the government of the People's Republic as the sole legitimate government of China and recognized Taiwan as part of its territory. At the same time, Japan lost faith in the United States but improved their relationship with China and the Soviet Union.[37] What is it that drives a state to promote its national interests by expanding relations beyond the allied bloc without breaking relations with other members of its own alliance? In order to gain insights into state processes and the motivations that underlie them, this book will explore the shifting sands of alliance cohesion which can sometimes generate a space in which a host state can expand its autonomy by looking after its own national interests. The book will focus on the role of diaspora communities, which can function either as a subset of a national homeland or as an 'imagined community' in transnational space, existing geographically outside the homeland and living in the hearts and minds of their people in

INTRODUCTION

the form of identity.[38] The diaspora's role can be expanded, contracted or suspended to reflect the policies of the host state, but broadly the host state will treat the diaspora under the policies that according to Mylonas can be categorized under three broad headings: accommodation, assimilation or exclusion.[39] These three categories of state policy will provide the lens through which changes in Japanese policy making attitudes towards the long-term Korean diaspora will be examined in response to geopolitical power configurations. The re-configuration of power over time has helped shape the varying degrees of alliance cohesion that have existed between the US, Japan and South Korea from the end of World War II until the present.

This introductory chapter will look at existing literature on the Zainichi in order to identify how academic scholarship explores the competing aspects of Zainichi identity. Existing literature offers perfectly good insights into the history and origin of the Zainichi, but fails to shed any light on the areas on which this book wishes to concentrate – namely the Zainichi's involvement in transnational space. On a broader level, the book will offer an overview on current diaspora literature that discusses the relationship between diasporas and the states of host country and homeland to discover why some ethnic groups become integrated into the host country while others do not. As far as the Zainichi are concerned, they represent a diaspora produced in the turbulent wake of World War II, during Japan's second nation-building process. Japan was a defeated nation, trying to find its feet in a radically changed postwar world. No longer an empire, the country was trying to rebuild itself along single-ethnicity nation-building concepts. This book will explore the multitude of approaches to nation-building processes in the postwar period and beyond by reviewing existing literature in the light of changing definitions of nationalism. Nationalism and its approaches are key factors underpinning a nation state, and their exploration will help explain how a diaspora is situated within the nation-building process. By incorporating a role for diasporas into mainstream International Relations theory, the work will shed a light on the driving mechanisms of Japanese state policies towards the Zainichi diaspora – policies which will ultimately produce a particular type of diasporic identity that belongs neither to host state nor homeland.

Korean Migration in Japan: Becoming Zainichi

Current literature on the Zainichi describes only one piece of a jigsaw that's been put together from various aspects of the diaspora within the domestic context of Japan. The Zainichi are characterized by numerous conflicting and contested notions of identity. Pre-existing complexities can be observed by studying academic debates on powerful binaries such as Korea and Japan, North and South Korea and, more generally, the colonizer and the colonized, as well as looking at different prevailing attitudes within younger and older generations. David Chapman focuses on such complexities within the Zainichi identity, explaining that its resistance to Japanese notions of national identity has often been framed through the binary logic of Japanese and Korean ethnicity and nationality. According to Chapman, struggles in defining the Zainichi identity based on such dichotomies often mirror the same homogenizing notions of Japanese identity.[40] The true nature of freedom from marginalization and oppression by the Japanese government's homogenization policy has also been affected by examining issues of essentialism within the Zainichi identity itself.

Against the essentialist aspects of the first generation of Zainichi diaspora, Kim Tong Myung looked at the diverse identification of contemporary Zainichi by focusing on the generation gap. Contemporary diaspora members are third or fourth generation immigrants and differences in approach, thought and outlook from one generation to the next have created diverse patterns within the diaspora. Kim argues that by clinging to the perpetual expectation of imminent reunification, the first generation of immigrants existed as 'provisional Zainichi', refusing to allow themselves any feelings of belonging or permanency.[41] Koreans referred to themselves as 'the Zainichi', a term that translates as 'staying in Japan', which shows that they perceived life in Japan as temporary. Kim, however, argues that it is now no longer practical for many Zainichi to return to Korea, since the basis of their lives (socially, economically and linguistically) has now become firmly rooted in Japan. Kim therefore proposes a 'third way', an approach that entails living in Japan under a Korean nationality, which would allow the Zainichi a sense of belonging to Japan without giving up their Korean identity.[42] Kim's 'third way' celebrates heterogeneity or hybridity, and challenges the rigid essentialism of both the older Zainichi identity and the postwar ideology of Japanese nation-building. Indeed, subsequent generations of Japanese-born Zainichi came resemble a cultural type closely related to Stuart Hall's definition – people whose presumed homeland has become 'an imagined community.'[43] The original 'Korea' is no longer there; in the same way that the original 'Africa' Hall describes has been transformed. Hall argues that 'diaspora', as a defining term, does not simply refer to scattered tribes whose identity can be related solely to some sacred homeland.[44] The diaspora experience is not defined by definitions of purity, but by the recognition of heterogeneity or hybridity.[45] While many cultural studies of diaspora are related to post-colonial theory, dominant nationalist discourses try to deny 'hybridity' in order to construct a homogeneous national space or identity. However, diasporic subjects embody the hybrid history of a nation – and indeed often seek to emphasize it by highlighting their cultural differences and retaining or amplifying their own cultural traditions.[46] In this context, diasporas serve as an important post-colonial critique.

Apichai Shipper, however, argues that the Zainichi diaspora's absence of loyalty to the host country is an obstacle to promoting a multicultural democratic civil society. Shipper's critique focuses mainly on the activity of the Korean ethnic associations in Japan, which are working with their overseas partners in the Korean homeland. Shipper criticizes the essentialist analyses of Zainichi Koreans by pointing out that they and their ethnic associations (*Mindan* and *Chongryun*) have little interest in improving living conditions for fellow co-ethnics or co-nationals within Japan. Shipper contends that former colonial subjects such as the Zainichi Koreans foster socioeconomic and cultural integration in some ways, but he argues that they also hinder national unity and deliberative democracy through their preoccupation with their homelands. They promote ethnic and class exclusivity and express disinterest in public discourse to improve conditions for fellow co-ethnics or co-nationals because they do not want to jeopardize their own privileged status by assisting newly arrived immigrants.[47]

The *Chongryun* (pro-North) and *Mindan* (pro-South) groups both emerged from the League of Koreans in Japan (*Choren*), an organization that was established by Korean nationalists and activists on 15 October 1945.[48] As *Choren* gradually fell under the control of Korean communists, a conservative group including former *Choren* members, established *Mindan* with Pak Yol as its first leader on 3 October, 1946. With the enactment of the South Korean nationality law in 1948, *Mindan* became the official organ of South Korea.[49] After the dissolution of *Choren* in 1949, those who had been active in the communist movement formed the Democratic Front for the Unification of Koreans in Japan (*Minsen*) to protest the Korean War and push for a democratic and communist unification of Korea.[50] Later, the North Korean regime came to view *Minsen'* s preoccupation with revolution in Japan as a mistake, and suggested that *Minsen* members loosen their ties with the Japan Communist Party (JCO) and concentrate on Korean problems. *Minsen* was dissolved on 26 May 1955.[51] One day later, left-wing labour activist Han Deok Su established the *Chongryun* under the direct control of North Korea. The Zainichi thus found themselves under the leadership of two insular organizations that encouraged Koreans to maintain their precarious status as foreigners with limited rights, asking them to ally themselves to one of two polarized and competing motherlands which had very different interstate relations with the host state. As a result, the Zainichi were condemned to remain a severely deprived and marginalized minority for at least the first part of the Cold War era.[52]

By focusing on Shipper's analysis of ethnic associations and their connections with the homeland, Sonia Ryang explored the vital mechanisms used by the ethnic organization claiming to represent members of the Zainichi diaspora who are ideologically connected to North Korea – the General Association of Korean Residents in Japan, known by its Korean name *Chongryun*. Ryang's work, *North Koreans in Japan*, concentrates on the *Chongryun* as an organization made up by members of a community who regard themselves not as an ethnic minority in Japan but as overseas nationals of North Korea. Apart from Ryang's work, the *Chongryun* has never been the subject of serious academic research, due to its isolation from larger Japanese society. The North Korean government encourages this isolation to promote the loyalty of Zainichi Koreans to North Korea, and *Chongryun* has been categorized as a highly disciplined Korean Communist organization working within Japan.[53] By taking advantage of her own ethnic background as a member of this community, Ryang sheds a light on the organization's internal workings.

Ryang's contribution helps explain how the external structure and the internal mechanisms of *Chongryun* helped promote North Korea's overseas

INTRODUCTION

influence by underscoring the processes by which individuals identify themselves as North Koreans.[54] But according to Ryang, the North Korean nationality ascribed to *Chongryun* Koreans is entirely manufactured and exists only as a symbolic identity. Although Pyongyang declared all Koreans in Japan as citizens of the DPRK, no consular services were ever provided and no passports have ever been issued. *Chongryun* Koreans have never voted in North Korean elections, nor have they been conscripted into the North Korean army.[55] However, the term 'overseas nationals of North Korea,' a term central to *Chongryun*'s legitimizing discourse, has helped forge a unique identity that has effectively replaced that of the colonial subject. In addition, despite the fact that the majority of Zainichi Koreans can trace their roots back to the southern regions of Korea, later to become South Korea, they made a choice to become North Korean nationals – even though that choice might result in precluding them from returning to South Korea, to their traditional provinces where their ancestors once lived.

Ryang argues that the *Chongryun* has always existed to feed on the complex interests and relations between the two Koreas and Japan in the context of Cold War in East Asia.[56] Since Ryang's central concern is the pro-North Korean diaspora's identity and its reproduction, she also explores the ways in which internal Japanese policies have severely divided the Zainichi diaspora in response to Cold War geopolitics. Ryang points out that the 1965 treaty between Japan and South Korea granted permanent residence status only to South Korean citizens in Japan, a status that was not made available to pro-North Korean residents.[57] However, Ryang does not explain how or why Japanese policy making toward the Zainichi Koreans, especially pro-North Korean residents, shifted so frequently from exclusion to accommodation. The approach Japan took towards the treatment of Korean nationals identified as supporters of the DPRK during the Cold War era – particularly during the period of normalization between Japan and Communist China (1971–2) – was characterized by a range of policies towards the pro-North Korean group which could be categorized as accommodation. During this period, for the first time since the end of World War II, pro-North Korean residents were issued with re-entry visas, allowing them to visit North Korea for family reunions even when no diplomatic relations existed between Japan and North Korea. There is

13

little literature exploring the role of the Zainichi from an international perspective, leading to a need to examine the conditions under which Japan employed policies inclusively or exclusively towards the diaspora.

The Relationship between Diaspora and Nation State

Diasporas can be considered as vital factors of modern nationalism, as critical constituents of nation statehood and as aspects of post-colonial geopolitics that have assumed greater levels of importance to more recent studies of political, social and international relationships. But despite modern contexts, the term 'diaspora' is not a new word. The term stems from the Greek διασπορά, which means 'scattering seeds', and was originally used with reference to the dispersal of the Jews from their ancestral homeland. The concept was also used to refer to Armenian and Greek communities. The Jews, the Armenians and the Greeks were categorized as archetypal diasporas, referenced in the context of the historical importance of the Jewish experience of exile, dispersal, and the promise of eventual return. The word has hitherto entailed not only a collective and continuing definition of a group whose alliance towards a homeland continued despite global dispersal, but carries with it the idea of re-unity in a new or re-imagined nation state, reflecting the Jewish experience.[58]

However, given the recent rise of transnational diaspora communities in the sphere of contemporary international geopolitics, some scholars argue that it is important to transcend the framework of the Jewish experience and explore diasporic concepts in a new way. James Clifford argues that by Safran's strict definition, the dispersed African, Caribbean, or South Asian populations do not qualify as diasporas because they are not as strongly oriented towards roots that exist in a specific place, nor do they express a desire to return. Clifford insists that a shared and ongoing history of displacement, suffering, adaptation or resistance is as important to the definition of diasporas as the projection of specific origins.[59] According to Khachig Tölöyan, the term diaspora, once associated with exile, dislocation, powerlessness and suffering, has become a useful and even desirable umbrella term to describe a range of dispersals.[60] Recent scholars of diaspora studies insist that it is more desirable to engage with the dynamics of different kinds of community identities or newly important actors in global politics than it is to focus on the debate of what groups truly deserve to be considered as diaspora. The processes of globalization offer transnational diaspora communities easy access to resources and strategies that may come to play significant roles in affecting politics in their home countries. The technological revolution has allowed for easy communication and the rapid transfer of information and even money from the diaspora communities to their families in the homeland, as well as linking kinship communities in other countries.[61] Arguably though, the host state is no longer viewed as the sole actor and diasporas have become the paradigmatic 'other' in the nation state, standing as manifestations of 'de-territorialized communities.'[62] Contemporary diaspora scholars are therefore starting to focus on how diasporas challenge traditional definitions of state or sovereignty because of the positions they occupy, taking an unbounded space between homeland and host country.[63] Since the collapse of the Soviet Union and the emergence of new multi-ethnic states, diasporas have become a much more important factor in the nation-building process. For example, separatist movements developed more quickly and more radically in the Baltic republics and South Caucasia than elsewhere in the USSR,[64] and the key factor for the comparative rapidity of these developments was that many new states had been established with the cooperation of their respective diaspora communities in the US and other Western countries.[65]

Yet what promotes the ethno-national consciousness of the diaspora? What drives diasporas to continue distinguishing themselves via their identity and resisting assimilation into the ruling nationality of the host state, allowing themselves to become defined as groups that are deeply engaged in ethnic mobilization in a transnational space? Brubaker explores the idea that an ethnic minority, a host state majority, and the minority's lobby state or homeland co-exist in a 'triadic' political space. According to Brubaker, dominant elites in the host state promote the state-bearing nation's language, culture, demographic position, economic growth and political hegemony, while the leaders of substantial, self-conscious, organized and politically alienated national minorities within those states demand cultural autonomy and resist processes of both assimilation and discrimination.[66] Brubaker goes on to say that the elites of the external national 'homeland' protect their diasporas against alleged violations of their rights and assert their rights – their obligations in fact – to defend the interests of their distant minorities. The triadic nexus theory's most important argument is that each of the elements in the relational nexus – minority, nationalizing state, and homeland – should be understood in relational terms, in which each political elite struggles for dominance within a range of competing stances.[67] But because the relational fields in the nexus are mutually constituted, Brubaker's theory cannot explain the ways in which fluctuations of ethnic mobilization shift over time.

Complicated cases such as the Zainichi diaspora represent a further challenge – their homeland is divided and the two halves take a competing stance against each other, so that the national minority, the host state majority and the minority's divided homeland lobbies turn Brubaker's triadic political space into a quadrilateral political space. The key point in the Zainichi's case is that the diaspora's dual loyalties to its divided homeland often give more leverage to the political elite of the host state than to each counterpart of the diaspora's homeland. For example, in contrast to the Koreans, the Taiwanese are said to have retained a fairly positive image of Japanese colonial rule – mostly due to their antipathy toward Chinese rule. Indeed, China has played an important role in the Taiwanese group's contradictory, conflicting identity contestation in Japan.[68] As such it becomes evident that diasporas can often display a conflicting agenda, finding themselves torn between a strong desire to retain their homeland's distinctive national characteristics and an equally strong desire to integrate into the host community. The conflicting agenda in the Zainichi's case has been created by the ambiguous policies of the host state, which have fluctuated between national interest and external security rather than considering the diaspora's own aspirations. Existing diaspora literature focuses mainly on the triangular relationships between homeland, host country and diaspora,[69] and does not take adequate account of the special case of a divided homeland.

It is also vital to understand the ways in which host state elites' policies towards diaspora groups change drastically from inclusive into exclusionary approaches or vice versa at any sudden changes in the geopolitical landscape. For example, history shows that Armenian diasporas were accommodated by the Ottoman Empire's approach under the heavily institutionalized *millet* (religious community) system.[70] But by 1875, when a new political approach had replaced the old one, the Armenian diaspora suddenly became the target of exclusionary policies, including genocide. Why was such a peacefully co-existing group suddenly treated so very differently? Mylonas argues that during the heyday of the Westphalia system, the overwhelming ethos of nationalism forced Ottoman Empire elites to replace the accommodation policy with one of enforced homogenization.[71] The history of the Zainichi diaspora shows a similar trajectory. During the colonial era, educational policies worked as an apparatus of homogenization in Korea, teaching Koreans to become subjects of the Emperor – a policy known as *kominka* (Imperialization) – by inducing them to accept the worship of shrines, to use the Japanese language and to take Japanese names. But this still does not explain why Koreans in Japan, as the imperial subjects of the Japanese Empire, suddenly found themselves re-categorized as foreigners at the end of World War II.

These questions encourage us to seek answers by investigating state policies on national integration. The study of nationalism is closely related to the formation of nation states, based as it is on national integration policies. Literature on nationalism differs in content and approach depending on various theoretical arguments, but is closely related to nation-building processes in terms of economic or territorial expansion, the establishment of political sovereignty and social and cultural behavioural norms.[72] In the process of nation-building, particular ethnic groups might be omitted from the mainstream narrative or integrated into the core population because they are considered either as integral to the policies of ruling elites or as objects for political manipulation.

Primordialist Perspectives on Nation

Primordialist perspectives strongly emphasize a nationalist consciousness which encompasses the idea of a collective identity based on shared claims to blood, soil or language. Power comes from myths, memories, traditions and symbols of ethnic heritage and the ways in which a nation's common past has been or can be rediscovered and reinterpreted from modern nationalist perspectives.[73] In other words, primordialists rely on studying widely diffused ethnic elements that recur throughout recorded history, such as ethnic origin myths, beliefs in ethnic election, the development of ethno-scopes, the territorialization of memory and the vernacular mobilization of communities.[74] Despite cultural fractures often introduced by conquest, colonization, migration or assimilation, ethnic groups endowed with a series of foundation myths inevitably become nations defined by nationalism, and the myths that are seen as part of that nation's eternal and unshakable destiny. Under primordialist agendas, nationalists find it impossible to integrate the diaspora community within the process of integration into the nation state.

According to Fiona Adamson, diasporas are distinct from other transnational groups (such as transnational economic networks or transnational advocacy networks) insofar as the networks of diaspora communities are defined as shared collective identities under ethnic, national or religious terms. Diasporas develop a particular identity even if it underpins a universal ideology such as nationalism.[75] The essentialism of the Zainichi identity, as some Zainichi literature argues[76], represents a typical diasporic element. From the view of the Japanese ruling elites, marking national boundaries and shaping identities were vital to the process of homogenizing notions of Japanese identity. Inevitably, the Koreans came to represent not only an exogenous other, but also an indigenous threat to the Japanese nation, the common 'us.' It might therefore have been reasonable for the ruling elites to employ either exclusionary or strict assimilation policies towards former colonial Koreans at the end of World War II.

The competing notions of the Zainichi diaspora identity are not simply categorized by ethnic or national issues. There are numerous other conflicting and contested notions of identity implied by the Zainichi diaspora, such as 'North or South Korea' and 'colonizer and colonized.' The category of 'North or South Korea' is not taken into account from the primordialist perspective, and the relation between colonizer and colonized implies historical antagonism, which may be a powerful identifier that helps create boundaries between nationalists and particular ethnic groups. But in the Zainichi's case, the situation is complicated further by the 'North or South Korea' category. Even though historical antagonism exists between Japan and South Korea, their split is not totally exclusionary and larger-scale geopolitics has compelled them to consult with one another over security concerns – especially regarding the threat posed by North Korea. Japan and South Korea have been embedded in the

INTRODUCTION

US-dominated East Asian geopolitical strategy as allies of the US, which has worked to exclude the USSR, mainland China and North Korea from normal commercial and diplomatic intercourse with the non-communist East Asian bloc during the postwar period and the Cold War. Security concerns have been prioritized above historical differences between Japan and South Korea. Korea's 38th parallel has for decades epitomized the frontier between capitalism and communism, maintaining and rigidifying the polarization of the two ideologies. The Zainichi, whose first generations came to Japan from an undivided nation, now find themselves compelled to choose a side not only by their host country but by the competing organizations within that host country which are striving for dominance to represent them.

The external national homelands of North and South Korea have for decades competed to redefine those ethnic Koreans who remained in Japan as their own nationals, discouraging their members from acquiring Japanese nationality throughout the Cold War.[77] For the Zainichi, political ideology became an increasingly important way of uniting disparate and dispersed ethnic populations within Japan. Rather than sticking to primordialist categories characterized by historical hatred between colonizers and colonized, the collective identity of the Zainichi instead replaced ethnic symbolism with political ideology. Adamson argues that political ideologies and identities are deployed as a means to unite disparate networks and groups into coherent transnational identity networks.[78] For the Zainichi, political identities and ideologies served as the cement that held the dispersed Korean community in Japan together during the height of the Cold War, which inevitably thrust the Zainichi onto the stage of world politics.

However, the primordialist approach cannot explain why some ethnic communities do not follow nationalist mobilizations, despite having the same mythical and cultural attributes and possessing strong feelings of belonging to the emerging nation. Unlike the Zainichi diaspora, the Sakhalin Koreans, who had been mobilized from Korea to Sakhalin during Japan's colonization era, did not continue to hold sympathy with North Korean leader Kim Il Sung. The North Korean association in Japan, the *Chongryun*, succeeded in persuading sympathetic members of the Zainichi diaspora to repatriate to North Korea, and for two or three decades this Korean diaspora remained surprisingly loyal to Kim Il Sung's regime. The *Chongryun* became a powerful agency within Japan, providing the North Korean regime with considerable funds and valuable intelligence at the expense of the host state. For a short while it seemed that Sakhalin Koreans would also side with Pyongyang, but this did not last, and Sakhalin Koreans soon grew disillusioned with North Korea. Although the majority of the Sakhalin Koreans were natives of the provinces of southern Korea – just as most of the Zainichi diaspora were – the contrast between the Sakhalin and the Zainichi comes down to a problem of allegiance; of a sense of belonging to the host country. As the Japanese economy boomed in the postwar period and standards of living improved dramatically, stories about life in North Korea began to filter out during the late 1950s and the briefly attractive mirage of North Korea very quickly collapsed.[79] Perhaps one answer to the question of belonging is that ethnic groups comply with host state policies due to economic interests at the expense of protecting their identity.[80] The Sakhalin Koreans felt it was more important to ensure their survival by integrating into the national society of the Soviet Union. In comparison with North Korea, the Soviet Union was a much more advanced country in the 1950s, a fact that was common knowledge among its allies.

Neither the endurance of ethno-national communities nor the power of ethnic myth helps explain why a particular set of circumstances encourages nationalism within some ethnic communities rather than across the entire ethnic spectrum. Political ties appear more important in establishing collective identity than any pre-existing ethnic bond. The process of identification takes place as a result of conscious effort by decision makers or elites, and identity politics is a process characterized by a conscious and deliberate effort to control a political position by defining the relationship between identity and difference, or changing mutual understanding and others' perception of the self. Unlike the unconscious social process of identity formation, the politics of identity denote a deliberate strategic action with a specific purpose. As Wendt argues, the process of identification forms a continuum that leads from negative to positive. In the process of positive identification, the other is seen as a cognitive extension of the self, while the idea of the other as anathema to self-results from that negative identification.[81] To be South Korean means *not* to be communist; to be North Korean means *not* to be capitalist or imperialist.[82]

INTRODUCTION

Constructivist Views on Nation

According to Brass, the cultural forms, values and practices of ethnic groups become political resources for elites competing for political power and economic advantage. Brass asserts that elites mobilize ethnic identities, simplify and distort their beliefs and values and then select those which are politically useful. In the first part of this process language, religion, familiar places and historical sites serve as identifying symbols on which the elites are able to build simplified and essentially spurious ethnic communities.[83] Elites seek to promote alternative ethnicities either as frightening negatives – as exemplified by the way Muslims are sometimes portrayed in the UK[84] – or as patronizing parodies of ethnicity such as the bogus ethnic sensibilities assigned to Native American communities. Either way, the elites present constructed and simplified ethnic groups to satisfy their own political agendas – usually to underscore their own intimations of nationalism.

The vital point in this part of the discussion is that the richness of cultural heritage from the point of view of the diaspora itself is not sufficient reason for the retention of ethnic solidarity over time. Before World War I, for example, the majority of Hungarian Jews knew no Yiddish at all and took no particular pride in developing any kind of Jewish ethos or strict religious practice; they chose instead to achieve economic success, establish social satisfaction and take Magyar nationality without religious conversion.[85] On the other hand, in some cases even the absence of joint cultural heritage cannot prevent some ethnic communities from building solidarity. In the 1960s and 1970s, African-Americans in the United States successfully won their civil rights, which contributed to maintaining a sense of ethnic identity despite the loss of a distinctive language. Likewise, in Europe a number of fully fledged, standardized written languages – such as Latin, Anglo-Saxon, Provençal, Low German and Slavonic – have been incorporated into other languages over the centuries despite falling from use in their own right.[86] Such examples demonstrate that the process of creating communities from ethnic groups requires the selection by the elites of language, religious practice or historical symbols.

According to constructivist views, it is not the place of birth, kinship, mother tongue or native religion that predetermines successful ethnic and national mobilizations, but rather the interests of the controlling elite.[87] The symbols selected to exemplify ethnic differentiation depend on the interests of the elite group taking up the ethnic cause. In the process of developing a nation from an ethnic group, nationalist leaders may emphasize different symbols and promote alternative conceptions of group boundaries.[88] In such a situation, ethnic group members gradually shift their own boundaries, and may choose to educate their children in a language other than their mother tongue in order to differentiate themselves further from competing ethnic groups. This often happens during the bargaining process between ethnic communities and state authorities, when ethnic leaders and members agree to give up aspects of their culture or modify their prejudices. Ultimately, many people never think about their language at all, and attach no emotional significance to it.

In contrast, other ethnic groups may preserve their ethnic language and emphasize their ethnic identity or construct boundaries against what they perceive as encroaching nationalism by setting up schools and colleges, especially if they share a common homeland or region of birth. Schools and education are vital instruments for controlling an ethnic group. Whoever controls the schools determines whether or not an ethnic group will maintain or lose its cultural distinction and thereby render itself susceptible to political mobilization on ethnic grounds.[89] Such differences in attitude on the part of ethnic communities serve to intensify mutually antagonistic ethnic viewpoints when seen in terms of the elites' rational choice of the cultural symbols relating to their own interests.

Why do dominant state elites pursue cultural homogenization? Do authenticity and cultural heritage still matter? One of the reasons for homogenization is that groups can be more readily mobilized on the basis of specific appeals. When ethnic appeals are made, pre-existing communal and educational institutions of minority ethnic groups will provide an effective means of political mobilization. Such cultural homogenization offers a basis for the continued debate that primordial perspectives still matter in terms of ethnicity. This argument explains why Zainichi Koreans in Japan were forced to have their children receive compulsory Japanese education from the end of World War II under the tight grip of the Ministry of Education. The national language not only contributes to integrating school children into the Japanese nation state but also provides a decisive role in sustaining control to ensure the dominance of a particular representation of ethnicity. From the constructivist viewpoint, Fredrik Barth argues that ethnicity is the product of a social process rather than a cultural given; something that is made and remade rather than taken for granted; chosen by circumstance rather than attributed to birth.[90] Ronald Suny also argues that the nation need not have been primordialized historically, but may have become so over time, until the point is reached where primordial ethno-nations became the template for nations themselves.[91] Hugh Seton-Watson suggests that a nation exists when a significant number of people in a community consider themselves as a nation, or behave as if they have formed one.[92]

The debate on national integration explores the way state elites produce and promote nationalist discourse and how this is perceived by the ethnic minorities.[93] Some states, such as China and Canada, accommodate non-national ethnic populations by encouraging their ethnic language and culture and by granting autonomous regions that are officially bilingual, with government documents and street signs in different languages as well.[94] The accommodation policy toward ethnic Koreans in China was based on close political ties between China and North Korea, which allowed diasporic Koreans to be integrated into the host country as Chinese citizens of Korean nationality. Unlike the Zainichi diaspora, the term 'our country' (*uri nara*), which for most Koreans abroad normally refers to Korea, for the Korean Chinese refers to the People's Republic of China.[95] Despite being encouraged to maintain their ethnic identity, Koreans in China were expected to reserve their nationalist feelings for the Chinese state. Minority groups, however, were vulnerable to sudden political shifts in China, which often contained xenophobic elements directed against non-Han ethnicities. Like other minorities, Koreans fell under suspicion as local ethnic nationalists during the Anti-Rightist Campaign of the late 1950s. Constructivist and instrumentalist approaches can both explain the fact that ethnic identity provides an effective means of political mobilization, but these views cannot account for the times when elites suddenly appear to perceive ethnicities as either useful or harmful.

Modernization Vision on Nation-Building

The Meiji state of Japan reflected aspects of unification between state and nation that are further explored by Ernest Gellner, whose analysis is that nationalism creates nations where they did not previously exist.[96] In contrast to agricultural communities, the modern industrial society is defined by an egalitarianism that is itself a by-product of the social fluidity of those industrial societies. All members of the earlier society possess skills such as literacy, numeracy, a basic work ethos and some technical abilities, yet Gellner conceives modernity as something structured along social orders from earlier societies overlain by industrialism. A new social mode of communication is therefore necessary to cultivate and develop the skills of a new society, and it is the modern nation state that fulfils that role by providing the organizational apparatus to develop high culture based on the ability to speak and write a common language; this in turn promotes cultural homogenization and communicative integration. According to Gellner, one of the nationalism's functions is to try to match politics and culture, organizing them into large, collectively educated and culturally homogeneous units.

Up to the end of World War II, national education in Japan was designed to produce members of the Japanese Empire as 'babies of the Emperor,' entities born out of imperial rule with a strong loyalty to the imperial household and the state.[97] The process of ethnic mixing, however, was controlled by a strong belief in Japanese superiority based on the supposedly divine origins of their race. Despite the rhetoric of multi-ethnic nation-building, the actual environment of Japanese colonialism opposed the idea of any true integration of the Japanese with their colonial subjects. But, in Japan, the inequalities remained; the Japanese were the superior race, and its colonial peoples were considered inferior. Largely subordinate in position and treated under separate colonial laws, indigenous populations lacked any form of representation in the Japanese Diet and had no effective legislative bodies of their own.[98] Moves towards integration simply intensified resistance from the colonial subjects to the pressure of acculturation.

Suny argues that empire is inequitable rule over subordinate colonies, while nation state rule is – in theory if not always in practice – equal for all members of the nation. Not all multinational, multi-cultural or multi-religious states are necessarily empires according to Suny, but where distinctions remain and treatment is unequal, usually in areas that remain ethnically distinct, then the relationship continues to be imperial, and inequitable treatment involves forms of cultural or linguistic discrimination.[99] Consistent with Suny's analysis, the Japanese Meiji state constituted an empire rather than a multi-national state. Although Japanese government elites during the colonial period advocated the goal of nation-building by using propaganda such as 'Japan and Korea as One' and 'Harmony between the Five Races' in Manchuria, they still insisted on the legitimacy of Japanese initiative to lead the Asian races.[100] Michael Hechter notes that once the core succeeds in persuading the population of its expanding territory to accept the legitimacy of its central authority, then nation state-building has occurred, but if the population rejects or resists that authority, then the centre has succeeded only in creating an empire.[101] The Japanese elites' project for constructing a multi-ethnic nation was terminated due to their defeat in World War II, and that defeat occurred before it was possible to determine whether Japan had created a nation state or an empire.

Geopolitics, Making Nation-Building Policies and the Zainichi Diaspora

Geopolitical shifts often occur during drastic transition periods, or in the wake of wars or sudden economic crises which inevitably cause a political, economic and social rupture. Large areas of opportunity are generated at the point of geopolitical change – opportunities that political actors are likely to seize. When this happens, diasporic groups and ethnic identities become problematic and can be seen by the state's new elite as an attractive target. The ethnic identity that every diaspora possesses is historically specific, but its effect has more relevance to geopolitical conditions. The geopolitical discourse of any country will vary over time because of its internal identities of and global positioning and the identities of the state's ethnic minorities and diasporic communities will reflect this.[102] Along the way, powerful but selective cultural empathies will be mobilized by which the state will enhance the legitimacy of its emerging economic, social and political relationships, providing its subjects with a new sense of national purpose and an altered identity. Thus ethnic identities cannot exist in a void, and must be understood in relation to economic and political factors, especially in relation to the international environment within which they emerged and have been reproduced. How do we examine the ways in which ethnic identity has been reproduced as diasporic identity? In other words, how do states put the same ethnic population into a different conceptual frame?

Brubaker suggests that diasporas should be treated under the category of political projects that connect emigrant communities with their homelands. 'We should think of diaspora not in substantial terms as a bounded entity, but rather as an idiom, a stance, a claim,' he says,[103] arguing that those who consistently adopt a diasporic stance often constitute a small minority of the population, forming a group within a host nation that political or cultural experts refer to as a symbolic ethnicity.[104] For example, some of the more politically committed Armenians express a desire to refer to all dispersed Armenians as a single diaspora because their conjoined diasporic nature forms a category of practice which is more useful for their political aspirations. However, the group that is casually referred to as the Armenian diaspora is, in some countries such as the US, not very diasporic at all, and the majority of those who identify themselves as Armenians distance themselves from diasporic stances, from links to the homeland and from links to Armenians in other countries. In that sense, the concept of diaspora is not a bounded entity in itself, but rather a stance or a claim. As such, 'diaspora' is an umbrella term that can be used to formulate and reaffirm the identities and loyalties of a population. In terms of groupism, this definition of a diaspora is quite different from Safran's primordialist approach.[105] But it is also necessary to make sure whether dormant members are really dormant; in other words, has their assimilation and integration into the host society been complete and total in nature? This standpoint underlines the reason why the primordialist approach is still used to some extent when defining diasporas. If it wasn't, why would some agents or actors try to provoke such essentialisms of belonging as language, religion and culture? Indeed, as Brubaker argues, the 'diaspora' card is a useful one for political agents or actors to play when making claims, articulating projects, formulating expectations, mobilizing energies or appealing to loyalties.[106]

INTRODUCTION

So as far as the Zainichi are concerned, how has the Korean ethnic identity been transformed into a diasporic identity in relation to the international environment during the post-World War II period? Moving from a domestic focus on 'ethnic' or 'non-citizen' groups to a transnational focus on diaspora allows new perspectives to come into play that introduce and incorporate the larger geopolitical context. As the empire-dominated system of the early twentieth century swiftly tipped towards a system dominated by nation states after World War II, changes that had begun to undermine imperial international relations since before World War II accelerated rapidly.[107] In the post-1945 period, a wave of decolonization highlighted empires as antiquated forms of government that were justifiable only as transitory arrangements that might aid the development of full nation states.[108] One of the most striking international changes was a dramatic shift in norms, as signified by international law and by the evolution of such institutions as the United Nations. Political ideologies such as self-determination, non-intervention, and sovereign equality of states became governing principles of interstate relations with dramatic speed. John Darwin argues that each former colony, no matter how artificial in ethnic or historical terms, should be entitled to self-determination.[109] The other systemic global change was the political and economic dominance of the United States and the Soviet Union that took shape soon after 1945, awarding military primacy in the international order to two powers that were both rhetorically anti-colonial, despite their own imperial legacies.[110] The economic dominance of the United States after 1945, in particular, brought with it the liberalization of international trade and investment, which in turn reduced the advantages of empires as large-scale economic entities.

These international transformations emerged in tandem with equally significant changes that helped overturn the traditional imperial bargaining within metropolitan and colonial societies. In Japan's case, instead of negotiating with its former colonies itself, the US and allied occupying forces provided the means and the rationale for conducting ethnic un-mixing of these countries.[111] Colonial subjects such as Koreans, Chinese and Taiwanese were encouraged to leave,[112] and it was not taken for granted that a new national or ethnic minority would stay or move based on their lack of rootedness, their sense of homeland or the extent of their homogenization.

During the Cold War era America reversed the course of its geopolitical strategy towards Japan and Korea. Initially planning to de-fang and enfeeble Japan, Washington instead decided to reconstitute Japan as a strong anti-Soviet ally.[113] In this way, Cold War geopolitical strategies were prioritized over historical reconciliation. Although feelings of distrust and contempt constituted the starting point for Japanese–Korean interaction, the triangular relationship between the two countries and America did not develop in a linear direction.[114] Inevitably the fundamental bilateral relationship between Japan and South Korea fluctuated between conflict and cooperation during the Cold War, depending on how the United States provided security assurance to the two countries throughout the period.[115] To alleviate its own fears over security threats, Japan occasionally pursued policies aimed at obtaining equidistance from both Koreas.

Given such a drastic change of international environment, it seems reasonable to assume that the Zainichi diaspora card was a useful one for agents or actors – including both Japan and South or North Korea – to play when making claims or articulating various political projects based on their own national interests. Such an arbitrary attitude towards the diaspora allowed the Japanese government to shift its policy toward the Zainichi, often on an ad hoc basis, either including them within its statehood or excluding them from it. It wasn't just Japan that did this – South or North Korea themselves utilized the diaspora card as well, often to mobilize energies by appealing to Zainichi loyalties for their own national interests. Thus the identity of the Zainichi diaspora was gradually transferred from one based on ethnicity into something more politically based. The trajectory of how Japan has treated its post-colonial Korean population cannot be explained by simply considering differences in ethnic identity or the relationship between colonizers and colonized. The politicizing of the Zainichi therefore left it in the unfortunate position as a bargaining tool between Japan and the Koreas, as well as external global forces. All parties seemed happy to allow the diaspora to be used as a bargaining tool or a channel of influence in a much higher political game.

Beyond the basic logic of interstate relations between host state and homeland, the Zainichi diaspora case study shows how a diaspora's collective identity can be shaped and influenced by geopolitical power configurations that go beyond not only the interstate relations between host state and homeland but also major global and regional powers. As a useful political tool, a policy of temporary accommodation was initially provided for the Zainichi, but this was abruptly curtailed and replaced with a strict choice between assimilation and exclusion. The ambiguous identification of the post-colonial Zainichi Koreans in Japan proves that policies of integration employed by elite nation-builders cannot be explained just by constructivist or primordialist approaches. By exploring the factors behind such rapid changes in attitude and policy, this book addresses the question of precisely how a diaspora can become something far great than – and much different from – its original or pre-existing cultural entity. By incorporating a more structured role for diasporas into mainstream International Relations theory, this book offsets the ambiguities of the constructivist approach, which fail to account for when and how elites play either the 'ethnic card' or the 'diaspora card.' But the rational egoism of states in an anarchical system – something that forms part of the main argument of neo-realist theory – can pinpoint the moment when elites use the diaspora card as well as why and how they do so. This book therefore explores the inevitable trajectory of the process of diaspora construction, as driven by a state-centric interstate relations system, concluding that diaspora identities are becoming increasingly significant and will remain so until the concept of diaspora ceases to exist or remains un-mobilized as a distinct political entity.[116]

1

Alliance Cohesion, Diaspora and Nation-Building Policies

A host state's national integration policies towards a non-assimilated group occasionally generate complicated results. Nowhere is this made more apparent than by studying the post-colonial Zainichi Koreans. Why have the Zainichi – fourth or even fifth generation in some cases – still not been fully integrated into their host country? What conditions control the way host states treat particular diasporic groups such as the Zainichi?

By focusing on the dynamics of alliances and alignments in East Asia, this book looks at how a diaspora's ambiguous identity affects its host state's behaviour when both host and home states are classified as minor or middle powers[1] as opposed to major powers. Such classifications depend on state capacity, which is measured by the state's administrative, policing and military capabilities in terms of its ability to maintain a relative balance of power against external threat as well as its capacity to implement its own domestic policies.[2] Major powers are likely to establish an order that favours their own interests and those of their populations in a strongly enforced approximation of peace and tranquillity.[3] Within such a hegemony, minor and middle powers are limited to playing supporting roles. In contrast to Europe, where the balance of power is maintained by an aggregation of states to prevent dominance by any single one, East Asian history favours hegemony or empire as a more common approach.[4] The minor and middle powers of Northeast Asia, however, are not just the pawns of the superpowers; to some extent they are more independent actors. They are likely to find themselves in the middle rank of material capabilities, but they possess the ability to maintain a degree of distance from direct involvement in major conflicts, a degree of autonomy in relation to major powers, a commitment to order and a sense of security in interstate relations as well.[5] For minor and middle powers, the balance between security and autonomy allows them to explore different approaches to promoting their national interests by using soft power rather than coercive power. As far as group relationship dynamics are concerned, the causal nexus between alliance cohesion among asymmetrically aligned states and the diasporas living within their borders offers a ripe area for study. It displays, in microcosm, an important underlying function of the mechanism of nation-building policy shifts by host state ruling elites, affecting their attitudes towards diaspora. As such, the host state treatment of diasporic groups may offer vitally important clues to the larger scale nation-building policies and international attitudes of those middle power states.

Nation-Building Policies via Interstate Relations

Studies by Harris Mylonas (2012) showed that interstate relations between a non-core group's host state and its homeland affect perceptions of the host state's ruling elites toward that non-core group. The underlying motivation for policy making towards that non-core group will therefore depend on whether or not the non-core group's homeland (or the notional country to which the ethnic group attaches its essential belonging or identity) is part of the same alliance bloc as the host state. Mylonas takes his observations beyond existing academic explanations – which emphasize ethnic hatred between groups, the importance of different understandings of nationhood and the primary focus on kin states – to build a theory to highlight and explain the geostrategic conditions under which certain host state policies become more likely towards ethnic groups.[6] By building on interstate relations between states and external powers (as opposed to simply considering the relationships between ethnic groups and host states) Mylonas attempts to predict a series of nation-building policy options and their possible outcomes. To do this he explores the mechanisms by which host state ruling elites might direct their nation-building policies towards non-core ethnic groups. Mylonas selects two independent variables for ruling elites' policy choice towards non-core groups. One is the foreign policy goals of the host state (such as revisionism or maintaining the status quo) and the other is the interstate relations between the host state and the external powers that support non-core groups within that host state. These interstate relations are measured by whether the relationship between the host state and the non-core group's home country take the form of common alliance or rivalry.[7]

Mylonas defines a core-group as the 'ruling political elites' rather than the 'nationalizing state', since the term 'ruling political elites' does not imply specific policies such as nationalization.[8] More particularly, he refers to the ruling political elites of minor or middle powers as those who are supported or controlled by greater external powers,[9] since their aim is the survival of the state and the preservation of their own position in it. Mylonas also explains the difference between non-core groups and minorities. The concept of a minority often implies either legal status, recognition from the host state or the existence of a claim by non-core groups. However, non-core groups do not necessary follow national or ethnic loyalties. According to Mylonas, referring to non-core groups as minorities carries a wider range of assumptions – such as the assumption that non-core groups will be targeted by policies of passive integration or will strive to obtain minority status.[10] In this book I apply the term 'diaspora' rather than 'non-core group' to the Zainichi minority, since the Zainichi have already been recognized by the host state and have raised their claims to the Japanese government often between the end of World War II and the present. Mylonas' definition of a minority is close to my definition of a diaspora, which was outlined in the previous chapter.

Mylonas's theory goes on to explain that state elites employ three levels of policy towards diasporas when engaged in nation-building: accommodation, assimilation, or exclusion. Accommodation refers to a host state's policy of retaining a non-core group within the state by preserving that group's cultural specificities and institutionalizing its minority status. Assimilation covers policies aimed at persuading (or compelling) non-core groups to adopt the core group's culture and way of life, and exclusion is the policy of physically removing the targeted group through population exchange, deportation or even genocide.[11] According to Mylonas, these three alternative approaches are employed according to cost-benefit calculations between the elimination of security threats posed by non-core groups and the cost of the necessary resources for the host state to eliminate those threats. Mylonas proceeds by presenting four causal paths that will influence and may eventually control the three policy choices towards non-core groups. Firstly, assimilation is more likely if the group has no external support; secondly, accommodation is more likely if an ally of the host state is supporting that group; thirdly, exclusion will result if an enemy is supporting the non-core group and the host state has revisionist aims and fourthly, assimilation through internal colonization is the likely outcome if an enemy is supporting the non-core group but the host state favours the status quo.[12]

Mylonas's framework is helpful in studying the Zainichi diaspora, whose principal defining feature as far as Japan is concerned is their divided homeland rather than their ethnic identity. Whilst it is obvious that the Zainichi diaspora comes from Korea, the more important question for Japan is whether its members offer their allegiance to North or South Korea – is their homeland part of an enemy or an allied bloc? This is a question that the Zainichi find hard to answer, as they are descended from the first wave of Korean migrants who came from an undivided state. This book will explain when and how the Japanese elites' perception of the Zainichi diaspora change from a single group into two different ones, and what drove subsequent state policies toward the two separate diaspora groups.

Diaspora and Divided External Powers

Although Koreans who remained in Japan were supposed to be a liberated people – unlike the defeated Japanese themselves – the Japanese government perceived the Korean population in Japan as a group deeply affected by US reversals.[13] From the perspective of US occupation policy makers, the large Korean group in Japan were regarded as the most part unassimilable, as a source of dangerous friction with the Japanese and as representing a strong element of instability in the Far East.[14] This concept made it too difficult for the US to move towards instigating bilateral diplomatic relations with either Japan or South Korea. When the San Francisco Peace Treaty came into force on 28 April 1952, John Foster Dulles was serving as US foreign policy adviser and was assigned the special responsibility of opening treaty negotiations with Japan. Dulles suggested that South Korea should be represented at the treaty conference, but Japanese Prime Minister Yoshida rejected the suggestion, saying that most of Japan's Korean residents were Communists, and would acquire all the compensation and property rights accruing to allied nationals if South Korea became a signatory.[15] There was a clear gap between Dulles' view that Seoul should participate in the first steps towards normalizing relations between Japan and South Korea and Yoshida's counter-argument that Korea was not one of the Allied Powers and should not be included under any conditions.[16] Japan did not officially want to recognize either the US-backed government in southern Korea in August 1948 or the USSR-backed government in North Korea in September 1948. In fact, Japan wanted to avoid any involvement on the peninsula. Although Japan and South Korea were viewed in a similar light under the US security umbrella, the two countries had no interest in forming normal diplomatic ties with each other, a reluctance that underlined the fact that they remained unallied for two decades after the end of World War II. This situation greatly influenced the treatment of the Zainichi diaspora by the Japanese host state, whose elites regarded the Koreans in much the same was as Yoshida himself. After the end of World War II, the Japanese government implemented the Alien Registration Law, which stipulated in 1947 that Koreans should be regarded as aliens. This stipulation represented the first step of the Japanese government's plan to enforce the repatriation what they saw as former colonial Korean subjects, as the Zainichi were now in direct violation of the provisions of the new Law.[17] Ethnic identities such as 'colonizer or colonized' were being rapidly transmuted into 'national or non-national', a process which ultimately created a transnational space – a notional area that was more political than geographical in which ethnic minorities, host state and homeland could co-exist. However, it is hard to see how that transnational space would function for the Zainichi, as their homeland was now divided and both halves were claiming that the Koreans in Japan were their citizens using criteria based on descent.

The Zainichi's unique status as a national minority meant that Brubaker's triadic political space turns into a quadrilateral one comprising the Zainichi, the Japanese and the North and South Koreans. The ambiguous loyalties of the Zainichi to their divided homeland allowed the political elites of the host state a wider range of choices than each half of the diaspora's homeland possessed. In the event of an allied and enemy external state fighting for control over the same diasporic group, the host state is likely to be offered more choice as to how to treat that group.[18] Japan's plan to establish a repatriation agreement with North Korea was successfully implemented in line with the revisionist foreign policies of both Japan and North Korea[19], despite South Korean President Rhee's statement that all Koreans were its citizens and repatriating any of them to the section of the country temporarily under enemy occupation was out of the question.[20] As Shain argues, a diaspora must be united in its claims on the host state in order to exert an effective influence on state foreign policy.[21] Divided groups within the diaspora's own community supported by the a divided and mutually antagonistic homeland will offer opposing views about the directions of the homeland's foreign policy, which may effectively cancel each other out. Divisions within the diaspora itself also carry their own inherent weakness, resulting in the provision of more choices to the host state.[22] As a result, the host state will be able to follow nation-building policies that leave non-assimilated diaspora groups weakened, subjected to the whims of the host nation who can treat the diasporic population as integral to their own internal policies or as subjects for political manipulation in the host state's own foreign policy.

The important point here is that host states may sometimes use a diaspora whose ethnic kin resides in a neighbouring country – regardless of whether that neighbouring country is part of an enemy or allied bloc – to further its own expansionist or revisionist regional objectives, especially when there is a high degree of revisionism driving the policies of that host country.[23] When this happens, the host state might find itself having to admit the risk that the diaspora community may affect host state policies toward their state of origin. For example, when the Pakistani government armed Afghan refugees in Pakistan during the Cold War, it limited its own options in its dealings with the governments of Afghanistan and the USSR.[24] Arming diaspora populations may also lead to internal violence and even civil wars in weak states, as exemplified by Kurds, Sikhs, Armenians, Sri Lankan Tamils and Palestinians. Mylonas' fundamental argument is therefore that support from an ally does not necessarily aim to underscore secessionist goals, but rather that host state elites are erring on the side of caution by preparing for the worst.[25]

State Capacity and Variations of Power Configurations under Asymmetric Alliances

The argument in this book focuses more specifically on the state capacities of allied member states and the perception of the value of alliance itself. These are important factors when assessing the degree of alliance cohesion. Focusing on the differences between minor and major powers allows the argument to clarify the concept underlying asymmetric and symmetric alliances by using Stephen Walt's definition; alliances may be symmetrical or asymmetrical, depending on whether the members possess roughly equal capabilities and take on broadly identical commitments to each other.[26] Mylonas also recognizes two important dimensions within the dynamics of external support that could provide additional considerations to the nation-building policy, namely policy priorities and asymmetrical alliances.[27] For such additional considerations to apply for case studies, he explores the necessity to relax the assumption of asymmetrical alliances in cases where the commitments and goals of the allies diverge.

In comparison with the bilateral alliances described by Mylonas, however, this book focuses on alliance blocs and explains that the triangular alliance between Japan, South Korea and the US displays a particular power configuration within a framework of an asymmetric alliance. This asymmetric alliance sometimes generates a space in which bargaining deals can be struck among allied member states, since for them the strategic gains are different to the ones that exist between symmetrically allied states. The normalization treaty between Japan and the Republic of Korea (ROK) in 1965 exemplifies the way that the US–Japan–ROK alliance produced diverse interests among asymmetrically allied states, in which all parties strove to follow their own interests throughout the bargaining and deal-brokering process. Ultimately, the strong alliance cohesion among the parties to this triangular asymmetric alliance allowed Japan to accommodate allied South Korean nationals by providing legal status for Zainichi who applied for South Korean citizenship in Japan. Interestingly, when Japanese elites perceived the asymmetric alliance as an axis for achieving Japanese foreign policy goals, Japanese foreign policy preferences as regards the politics of the Korean peninsula fell from neutrality into a state of common alliance with US, following and supporting South Korea. This marked the turning point at which the perception of Japanese government toward the Zainichi diaspora changed from a single category of threat into two different political categories: one supported by an allied state and the other supported by enemy state. In terms of state capacity, the value of any multi-state alliance depends on the extent to which the host state can obtain benefits through bargaining deals within an asymmetric alliance framework. In other words, the deals can be bargained or brokered by a diaspora's host state with other powers within the alliance bloc – powers both big and small, and powers whose levels of influence within the alliance can very over time.

Regarding this deal-bargaining, I use the framework proposed by James Morrow, who has observed that opportunities to form symmetric alliances will be rare because such alliances require a significant harmony of interest. Each party is likely to gain similar benefits under a symmetric alliance, so the interests of both parties must be sufficiently close for them to form an alliance in the first place. States will gain security (if their interests in preserving the status quo match) or autonomy (if their interests in changing the status quo complement one another), as long as the cost to each other in forming an alliance is not too high.[28] However, if state interests are not aligned, an asymmetric alliance might offer an alternative policy, as long as the circumstances exist under which it becomes easy to maintain a mutual harmony of interest, since under an asymmetric alliance each side will receive different benefits by bargaining between security and autonomy.

Minor powers, according to Morrow, have lower levels of security and higher levels of autonomy and will therefore try to form alliances that increase their security at the cost of some autonomy. A major power can offer a potential ally a significant increase in security, but in return it will demand a high price in autonomy when the alliance is formed.[29] In wartime situations, or when war threatens, alliances generally involve some pre-war preparations, which usually centre on the coordination of military and foreign policy to reinforce campaign plans and to specialize their forces in readiness. Coordination of foreign policy eliminates conflicts of interest that might otherwise separate allies.[30] As Morrow argues, national interests can be represented by a spatial model;[31] if the space for security shrinks, the space for autonomy grows. So the tighter the alliance, the greater the coordination between them. Conversely, if alliance cohesion is weak, each allied state's space for autonomy will increase.

But different power configurations set in the framework of an asymmetric alliance will produce different patterns. For example, North Korea's position with regard to the USSR and China was different during the Cold War era, especially during times of Sino–Soviet conflict. North Korea became strategically important for both great powers, with China maintaining a rigid one-Korea policy and rebuffing Seoul's overtures during the Sino–American détente. In contrast, Japan and South Korea, whilst not allies, could at least be considered as existing in a state of alignment that depended on the cohesion of separate US–Japan and US–ROK alliances. Borrowing Victor Cha's quasi-alliance model, this book explores the ways in which different power configurations among asymmetrically allied states maintain alliance formation. In the case of the Japan–ROK–US alliance, a reluctance on the part of two minor states to treat each other with respect or pursue closer relations with each other stems from their consideration of each other as secondary to the primary relationship, the one with the major state. While both Japan and Korea have asymmetric fears of abandonment and entrapment as regards each other, they share an identical fear of the US abandoning its defence commitment to the region as a whole. In contrast, the USSR and China acted as major states competing for closer relations with the minor state, North Korea, in order to enhance their own strategic influence. In this case it was the major states that shared the fears about abandoning their defence of the region. As long as both shared the same value towards the minor state, they maintained their alliance with North Korea. State capacity within asymmetric alliances is important, since the alliances provide a space in which each state is able to conduct bargaining deals based on security and autonomy depending on its own strategic foreign policy. For minor powers, this offers a good opportunity to promote their own foreign policy – even when they lack any significant military capability.

The cohesion of alliance can be measured by using a few major concepts. In line with Glenn Snyder (1990), this book considers alignment or alliance to be a summary of the expectations held by policymakers in response to the questions: 'Who will defend whom?' or 'Who will support whom, and who will resist whom, and to what extent?' Expectations of support or opposition can stem from various sources: inequalities of strength, conflicts and common interests among states and past interactions. Policymakers' expectations will be weaker if the lines of conflict and commonality concern different issues, but will be stronger if connections are mutually reinforcing. In addition, aspects of conflict and commonality will combine with factors of inequality of strength to produce a pattern of alignment.[32] The main question – 'who will defend whom?' – seems to be more concerned with superpower alliances, but when exogenous geopolitical situations shift, lesser powers – especially if they think they will find themselves at the mercy of altered superpower alliances – will examine the changing geopolitical situation and try to predict 'Who will support whom, and who will resist whom, and to what extent?'

Diaspora as an Alternative to Nation State

Given these circumstances and the predicted behaviour of middle-power host states, the next question to ask is how will movements in global and international politics affect the host state and homeland's attitudes towards diasporas? Will they affect the way a host state views its diasporic population, and given the unique situation of a divided homeland, how will large-scale geopolitical changes affect the treatment of the Zainichi diaspora as a specific case study? State processes and motivations underpin the shifting structures of alliance cohesion, and movements within these structures can sometimes generate a space in which a host state is able to expand its autonomy by looking after its own national interests. The diaspora coexists in a space of its own, and serves as a link between international and domestic spheres of politics.

This book follows Brubaker's conceptualization of diaspora as a useful test case; as a category of practice. A diaspora is a useful group on which a host state can practice the use of flexible tools for increasing its autonomy. By accepting Brubaker's framework, I use a diaspora as a substitute for

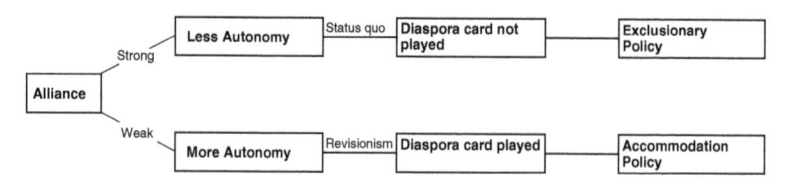

```
                              Status quo  ┌─────────────────┐       ┌──────────────┐
                    ┌─Less Autonomy──────│Diaspora card not│───────│ Exclusionary │
             Strong │                     │ played          │       │ Policy       │
┌────────┐          │                     └─────────────────┘       └──────────────┘
│Alliance│
└────────┘  Weak    │   Revisionism       ┌─────────────────┐       ┌──────────────┐
                    └─More Autonomy──────│Diaspora card    │───────│ Accommodation│
                                          │ played          │       │ Policy       │
                                          └─────────────────┘       └──────────────┘
```

Figure 1.1 Playing the diaspora card – the role of interstate alliances in the treatment of diaspora groups

nation states – as another way to essentialize the process of belonging and as a representation for the de-territorialization of identity. A diaspora is, in some cases, more flexible and useful than a nation state, which is defined as a rigidly bounded territorial community. A diaspora on the other hand represents a de-territorialized identity that has been reconfigured and stretched in space to cross state boundaries, although the connective presupposition is that this identity still remains on some level fundamentally consistent.[33] As such, viewing a diaspora as a category of practice is a useful exercise for both nation state and diaspora entrepreneurs, giving them a basis to make claims, articulate projects, formulate expectations, mobilize energies and appeal to loyalties.

But what if host states willingly play the diaspora card, using the diaspora as leverage rather than considering it a potential threat? Such a practice would enable host states themselves to emulate diasporic practices; to stretch in space across state boundaries in the same way diasporas are seen to do. Figure 1.1 shows the potential for using or ignoring the diaspora card, and indicates situations in which it could be used or ignored. As previously explained, when the critical exogenous geopolitical situation shifts, host states – especially if they are lesser powers who will find themselves at the mercy of altered superpower alliances – will consider 'Who will support whom, and who will resist whom, and to what extent?' It is at this juncture that the space occupied by the diaspora moves, either by expanding or contracting in relation to host state policies. Host states either accommodate or exclude the diaspora to further their own national interests by making a rational choice between security and autonomy within the framework of an asymmetric alliance.

According to Shain and Barth, diasporas represent some of the most prominent links between international and domestic spheres of politics.[34]

This observation is based on the interaction between host states and external powers on one hand, and host states and diasporas on the other. A weak state is susceptible to acute fears of abandonment, while a stronger one possesses the ability to balance itself internally against outside threats, giving it a greater range of choice and offering less intense fears of abandonment by its allies.[35] The presence of a diaspora is likely to compromise aspects of inequality between a weak state and a stronger one in terms of rational calculations regarding state-level trade-offs between autonomy and security. Policies towards a diaspora by the host state are made either inclusively or exclusively as a reflection of bargaining deals between autonomy and security that were struck under a framework of asymmetric alliance, which is itself a response to geopolitical power configurations. Thus diasporas occupy the middle ground between host state and homeland, and their role as an alternative to the nation state allows them to affect the behaviour of host states under conditions of asymmetric alliance.

External involvement vs Domestic Politics

There may be other factors that explain the Japanese government's fluctuating policy towards the Zainichi diaspora. These factors relate to domestic considerations rather than external international factors such as alliances. For example, Japanese public opinion towards South or North Korea might have influenced Japanese policies towards the Zainichi, but this cannot explain how the 1965 ROK–Japan Normalization Treaty was accomplished without the cohesive strategy of asymmetric alliance. Most Japanese intellectuals opposed the treaty, insisting that it would aggravate tensions between North and South Korea by compensating only the southern regime, which was at the time widely regarded as a military dictatorship. Japanese lawyers expressed their dissatisfaction by asking the Ministry of Justice why the granting of permanent residence was possible only after the signing of a treaty, since it was a matter the Japanese government could have solved simply through direct negotiation with Koreans in Japan.[36] Many leftist students and activists also waged strong campaigns against the treaty.[37] In early February 1965, grass-roots involvement in the campaign against the ROK–Japan Treaty began to gain momentum and spread throughout Japan.[38] Given that the opinion of the majority of Japanese was that the Zainichi Koreans had criminal tendencies, ordinary Japanese feared that many of them would become eligible to reside in Japan permanently. Despite the mobilization of the masses as well as public anti-Korean feeling, large-scale protests failed to block the ROK–Japan Normalization Treaty. In subsequent empirical chapters, I will expand on the way Japanese public opinions influence the government's fluctuating policy toward the Zainichi diaspora. I will also explore whether these policies can be fully explained by external fluctuations or by domestic considerations driven by Japanese public opinion.

Another aspect to be considered is whether the actions of the diasporic population and the factors that drove them was controlled by domestic politics or external influences. Despite their transnational location, diasporic activities and the influence they have in the diaspora's homeland have expanded the meaning of the term 'domestic politics' to include not only politics within the state but also within the population.[39] My argument is that external involvement is an inescapable aspect of Japanese policies toward Zainichi. After the end of World War II, there were leftist homegrown movements organizing Korean repatriation and establishing Korean language schools, which played a role in the nation-wide organization of Koreans in Japan. The first organization for Korean residents in Japan, the *Choren* (the League of Koreans in Japan) was formed shortly after Japan's surrender under the American occupation. Initially politically diverse, *Choren* became more left-wing over time, staffed by members of the Japanese Communist Party (JCP) who were politically radical and highly active in the Korean independence struggle. The veteran Korean Communist leader Kim Chong Hae exerted considerable influence on the group.[40] The occupation forces, in the face of the league's radical left leanings, moved against it and on 8 September 1949 the league was dissolved without preliminary warning by order of the Ministry of Justice. When the Korean War broke out less than a year later, former members of the radical group *Choren* established the underground Democratic Front for the Unification of Korea (*Minsen*) on 15 August 1950.[41] *Minsen*'s preoccupation with revolution in Japan, however, did not capture the popularity of the Zainichi diaspora but seemed to go against the North's revisionist plan to establish diplomatic relations with Japan. *Minsen* was dissolved in May 1955 when the DPRK announced its intentions of normalizing relations with Japan, and was replaced by *Chongryun* the day after its dissolution.

The diaspora category was a useful card for the ruling Japanese elites to play when devising state policies that might emphasize domestic politics in terms of the treatment of the Zainichi. However, the fact that these homegrown movements did not last long enough to root themselves into the diaspora shows that external factors such as the changing influence of their homeland were affecting the behavior of the Zainichi minority to a much greater extent than host state polices in Japan. As Erin Jenne argues, where diasporas are concerned it is ill-advised to limit one's scope of inquiry to the boundaries of either host state or homeland due to the increasing porosity of state boundaries and the increasing impact of external factors on domestic politics.[42] The Zainichi diaspora could not have survived without support from its divided homeland or the specific conditions that pertained to shifts in alliance cohesion.

Possible Causal Pathways

During the Cold War, the shifting balance between security and autonomy significantly affected the cohesion of the Japan–ROK alliance, influencing Japanese policies toward the Zainichi diaspora and causing frequent shifts. In order to re-configure the diaspora's space in response to a continually fluctuating exogenous balance of power, the essence of the diaspora has to stretch across state boundaries (even if the space it crosses is considered dangerous enemy space) and encompass future capabilities. Alliance cohesion is tested when the critical exogenous geopolitical situation shifts, at which point, as previously noted, states will consider: 'Who will support whom, and who will resist whom, and to what extent?' This is the point at which the space occupied by the diaspora moves, either by expanding or contracting to reflect the policies employed by the host state: exclusion, assimilation or accommodation.

This is where the theories laid down and explored in this book come into play. By combining the logic of alliance cohesion with Mylonas' nation-building theory (which explains interstate relations between homeland and host state), I conclude that these assumption can give rise to certain configurations which cover all eventualities of possible treatment of diasporic populations and summarize the positions in which most diasporas – including the Zainichi – eventually find themselves. These configurations can be summarized as follows:

Configuration 1: if alliance bloc cohesion is strong, the host state policy will be one of *exclusion* toward a diaspora group supported by an enemy state, but one of *inclusion* toward a diaspora group supported by an allied homeland.
Configuration 2: if alliance bloc cohesion is weak, the host state will *accommodate* a diaspora group supported by a homeland belonging to an enemy alliance bloc in order to expand its own autonomy.
Configuration 3: if the host state and homeland have no diplomatic relations, then the host state policy toward the diaspora group perceived to be backed an enemy homeland will be *exclusion* (See Figure 1.2).

Japan and South Korea became strong allies of the United States and, with both countries looking to America for security and protection, their relationship with each other was of secondary concern. The triangular asymmetric alliance combines two unallied minor states that share a major state as a common third party ally. But without diplomatic relations between the commonly allied minor states with the major state, alliance strategies

		External support		
		Interstate relations/**allied** bloc	Interstate relations/**enemy** bloc	**No diplomatic/ enemy relation**
The degree of alliance cohesion among asymmetric allied bloc	Alliance cohesion/ **strong**	Inclusion	Exclusion	Exclusion
	Alliance cohesion/ **weak**	Inclusion	Accommodation	

Figure 1.2 Causal pathways: alliance cohesion, interstate relations and policy toward the Zainichi
Note: Inclusion policy takes the form of assimilation, which is a different form of inclusive policy to accommodation.

and goals of those allied states cannot work effectively. Victor Cha calls the relationship between Japan and South Korea an 'alignment despite antagonism,'[43] yet without diplomatic relations, the relationship between the two states would simply be antagonism. If the external power which supports the diaspora group is perceived as an enemy by the host state government, that government is more likely to use repressive policies against that diaspora, leading to **Configuration 3**.

For the Zainichi, the problem is of course further complicated by the North/South division of its homeland. Japan and South Korea are in alliance with the US, while the North Koreans were being fought over by the Russians and the Chinese. The Zainichi, who initially came from an undivided homeland, find themselves caught in the middle, subjected to a tug-of-war for their allegiance between North and South Korean institutions in Japan who are struggling for the diaspora's allegiance rather than working towards its material benefit or security. Studying the Zainichi in this light can help understand the stresses and pressures put on any diaspora by the process of alliance cohesion I outlined above, and will prove invaluable to future studies on alliances, the treatment of diasporas and the psychology of national and ethnic identity.

In any alliance, if the cohesion is weak, the space for autonomy within that alliance will expand to reflect the fact that host state policies will protect the state's own national interests, including economic expansion and defence. Weak alliance cohesion will offer incentives to host states to politicize ethnic populations by utilizing the diaspora card in favour of their own revisionist aims. But the diaspora card can also assume negative characteristics under conditions of weak alliance. For example, from the homeland's viewpoint, a weak alliance cohesion increases the validity of ethnicity; this means that other factors such as colonial legacy and residual historical hatred can easily become heightened when trying to use either ethnicity or diaspora as a category of practice. But when allied countries face a threat from an enemy country, antagonistic historical relationships between host and homeland will be put aside, and they will try to make the best of their past relationships by cooperating in a foreign policy designed to retain and support harmonious interests by eliminating conflicts of interest that could separate them. Ironically, alliance provides incentives to de-politicize ethnic identity in both host country and the homeland in order to fight together more effectively against a common enemy at the expense of peacetime ethnic contingencies. The host state will treat the diaspora under policies that Mylonas identifies as accommodation, assimilation or exclusion. For the Zainichi, these configurations cover shifts in Japanese policy making attitudes towards the long-term Korean diaspora that have resulted from geopolitical power re-configurations offered by the varying degrees of alliance cohesion between the US, Japan and South Korea from the end of World War II until the present – an alliance that has been stretched, tightened, weakened and above all constantly tested by the presence of North Korea and its own alliances with different and conflicting superpowers.

Research Design

From the discussions above, I conclude that nation-building policies not only reflect ethnic choices or cultural differences within host nations, but are also driven by external factors. Pre-existing ties with an external homeland supported by allied or enemy groups – including an asymmetrical power relationship with the United States – is an independent variable, while policies followed towards ethnic groups or diasporas by host states is a dependent variable. The strength of alliance cohesion determines the host state's policy making – inclusively or exclusively in Japan's case towards the Zainichi diaspora, which has both an allied and an enemy external homeland (South or North Korea). Alliance cohesion is tested when critical exogenous geopolitical factors change, and the space occupied by the diaspora moves in consequence, either by expanding or contracting in relation to the policies employed by the host state. But given the divided loyalties of the Zainichi, how will that diasporic space move?

In order to evaluate the validity of the three configuration hypotheses delineated above in the specific case of the Zainichi, this book divides the decades since the end of World War II into three separate eras: 1945–64, the Cold War period from 1965 to the end of the 1980s, and the post-Cold War period from 1990 to the present.

I first consider the period from 1945 to 1964, during which all Koreans in Japan suddenly became refugees and were labeled enemy aliens because of the reversal of American occupation policies, which was itself due to the rising power of Communism in East Asia. These years were marked by the absence of any alliance between the homeland and the host country. But 1965 saw the signing of the Japan–ROK Normalization Treaty; a turning point for Japanese policy that ushered in a period when the Zainichi diaspora was recognized as being split between a common allied homeland and an enemy allied homeland.

The second era spans the Cold War period from 1965 to the late 1980s, an era that was characterized by a triangular alliance between the US, Japan and South Korea. In discussing this period, I borrow from Victor Cha's quasi alliance theory, which breaks the era down further into periods of 'cooperation' and 'friction' during the Cold War. Cha divides the Cold War era into four smaller time periods: 1965–71 (Cooperation under the Nixon Doctrine); 1972–4 (Détente and Heightening Crisis); 1975–9 (Post-Vietnam and the Carter Years); and the 1980s (Nordpolitik). Within these periods, the years 1965–71 and 1975–9 can be categorized as periods of cooperation, since both Japan and South Korea perceived and shared fears of abandonment regarding their primary security guarantor, the United States. On the other hand, the years 1972–4 and the 1980s were times of greater friction – a period that began with Japan normalizing relations with China and ended with Japan increased its military spending, expanding its sea lane defence perimeter to one thousand miles and revising its historical textbooks to reflect changing attitudes. The periods of cooperation were defined by a status quo foreign policy, while the frictional times featured revisionist foreign policies on behalf of the host country. During the periods of friction the historical colonial arguments became controversial issues.

In 1990 the Cold War ended and South Korea and the Soviet Union drew up a treaty of normalization, which was reflected in a logical change of treatment towards Japan's attitude to the Zainichi. The third postwar era thus begins with the end of the Cold War, when new geopolitical power configurations rendered the strategic balance in North Asia much more complex. As Brett Ashley Leeds and Burcu Savun explain, alliances are more likely to be abrogated on an ad hoc basis when one or more members experience changes that affect the value of that alliance or their perception of it, such as changes in international power, changes in domestic political institutions or the formation of new external alliances.[44] Leeds and Savun identify four factors crucial to the value of alliances – the level of external threat faced by the allies, the military capabilities of the allied states, the extent to which policy goals are shared by the allies and the availability of substitute allies. By focusing on these four factors, I sub-divide the post-Cold War years into the weak alliance cohesion years (1990–2), then the strong alliance cohesion years (1993–2002) followed by a return to a weak alliance cohesion from the autumn of 2002 to the present.

These divisions will enable me to test a set of hypotheses predicting possible policymaking toward the Zainichi diaspora by Japan as a reflection of the degree of alliance cohesion between host state and homeland in response to changes in geopolitical power configurations. When conditions change, as they did with the end of the Cold War, the host state may treat diaspora groups in different ways, and treatment will depend on additional factors that relate to the limitations of the correlation between alliance cohesion and diaspora as a category. Specifically, I examine the state capacities of common and enemy allied nations and their power configuration within the framework of alliance as well as regional geopolitical power competition. Firstly, if we consider enemy alliances, the Soviet Union was no longer a major power after the Cold War, and the perception of North Korea is thus subject to change as far as Chinese perspectives are concerned. China no longer needs to maintain a one-Korea policy to keep Russia at bay – indeed, in 1992, China made a normalization treaty with South Korea. Secondly, the dynamics of US power among asymmetric allied member states changed dramatically. During the Cold War era, minor states such as Japan and South Korea were both protected under the US security umbrella. In the post-Cold War era, however, the supersizing of the North Korean military threat become a major independent factor which could no longer be offset by bargaining between autonomy and security among allied member states. Thirdly, the rising state capacity of China and the relative decline of US power affected regional geopolitical power re-configurations. These three factors affected the behaviour of Japan and South Korea in a different way than during the Cold War era. Rather than playing the diaspora card indirectly, Japan and both Koreas had to directly reconstruct state capacity against different perceived external threats. North Korea attempted to increase its own military capability, while South Korea sided with the enemy state for its own security by signing a normalization treaty with China. As far as all countries involved were concerned, policy goals among allied member states will be more diverse under an asymmetric alliance. In that scenario, I can identify more precisely the conditions under which Japan will treat the Zainichi diaspora either inclusively or exclusively.

For the Cold War and the post-Cold War periods, I will test my hypotheses by using a process tracing method. Process-tracing has many advantages for theory development and theory testing, allowing me to check for false information and recognise causal influence on the basis of a few cases or even a single case.[45] Charles Tilly emphasized the importance of process-tracing, urging that theoretical propositions be based not on large-N statistical analysis but on 'relevant, verifiable causal stories resting in differing chains of cause-effect relations whose efficacy can be demonstrated independently of those stories'.[46] David Collier, however, argues that internal comparisons are critical to the validity of small-n analysis.[47] David Latin also emphasizes the importance of theoretically oriented process-tracing, which has made a fundamental contribution to finding regularities through the comparative juxtaposition of historical cases.[48] This book therefore uses a diachronic comparison method (comparing cases over time), which requires a set of cases (various combinations between alliance cohesion and policy choice) for which data exist at three known points in time or during the historical periods I identify as post-World War II, Cold War and post-Cold War, ultimately defining an initial and final state. Within the same study, it is possible to compare Japan's policy making towards the Zainichi over time to assess why some policies toward the diaspora were exclusive while others were inclusive. By establishing similarities and differences in the manner of policy change over time, I will be able to establish not only what happened but also how and why. By analysing empirical results against the hypotheses, I can identify the conditions under which state policies were employed either inclusively or exclusively, and by tracing the causal effects of these policies by the state, this book will be able to recommend a set of theoretical recommendations for effective integration policies to host state policy makers for the future.

Empirical evidence for the book was collected from archival materials,[49] daily newspapers[50] and government documents,[51] as well as from various secondary sources. Although some secondary sources are mainly

English or Korean documents, some material is of Japanese origin. Because the book considers policy making by the Japanese government, I examined Japanese newspapers covering the relevant period. As Mylonas defines core-groups as ruling elites, I also focus on elite discourses by examining newspapers and archival documents. Archival material is very useful since it describes in detail the interaction between US presidents and other influential policy makers and the Japanese ruling elites as well as South Korean presidents since the end of World War II. This helps clarify the real motivations underlying the Japanese ruling elites' selection of certain policies towards the Zainichi. By tracing these materials chronologically, the book identifies critical junctures and policy shifts towards the Zainichi by ruling elites that reflect geopolitical events. At the same time, process-tracing can capture the reactions of the Zainichi to these policy changes as well as how both divided external homelands reacted at these critical junctures. This will enable me to analyse the mechanisms behind the Japanese elites' policy shifts towards the Zainichi diaspora.

2

The Zainichi Diaspora: From the Shadow of Japan's Colonial Legacy

Introduction

Who are the ethnic Koreans that Japanese society identifies as the Zainichi? The answer isn't straightforward. Although most people consider them to be Korean-Japanese who live in Japan and possess Japanese nationality, naturalized Koreans may be many in number but they are no longer considered true Zainichi.[1] From the legal perspective, the Zainichi are defined firstly as Korean nationals who moved to Japan before the annexation, secondly as people who moved to the Japanese territories during the colonial period and thirdly as Korean nationals who remained in Japan after the end of World War II.[2] In historical terms, without colonization during the final years of Japan's empire and the country's subsequent defeat in World War II, there would be no such thing as a Zainichi Korean minority in Japan.

According to *Teikoku Tokei Nenkan* (the Statistical Yearbook of the Empire), there were only four Koreans in Japan in 1882. By 1909, the year before the annexation, the number had grown to 790.[3] Prior to 1910 the Korean community in Japan had consisted of students, political exiles and consular officials. However, as early industrialization began to define the Meiji era, the Japanese economy started to introduce a Korean workforce. Michael Weiner explains that the main causes for Korean migration to

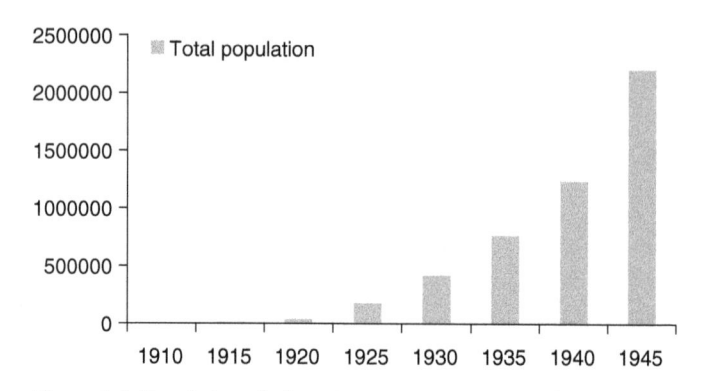

Figure 2.1 Population of ethnic Koreans in Japan, according to the census, 1910–45
Source: Toshiyuki Tamura, 1999. *Korean Population in Japan 1910–1945.* (1)–(3) Keizai-to Keizaigakusha.

Japan were twofold: the first was colonial policies, which led to the rapid growth of a large, impoverished and landless peasant class in southern Korea and a failure to develop an industrial base there which could have absorbed a considerable proportion of excess rural labour those policies created; the second was the instability of an essentially cheap labour market in Japan which could not meet the demands of wartime industry.[4] When Japan embarked on the conquest of China and eventually declared war on the United States and its allies, Korea was to serve as the forward logistical base for Japanese expansion, together with Manchuria in the north. At the same time, Korea was ordered to produce manpower to supplement the Japanese workforce.[5] By the end of World War II, Japan had mobilized approximately two million Koreans for hard labour in Japan, Sakhalin and the South Pacific.[6] (See Figure 2.1)

It is extremely significant that the history of Japanese colonization is inextricably linked with the creation of a Korean diaspora. There are currently 5.7 million Koreans scattered across 151 countries, although most of them are concentrated within five countries: the United States (2 million), China (1.9 million), Japan (0.6 million) and the Commonwealth of Independent States or CIS (0.5 million). These five countries account for more than 90 per cent of overseas Koreans[7] and, with the exception of the US, they are all related to Japanese imperial expansion. During

Japan's colonization period, many Koreans (mainly from what is now North Korea) migrated to China and settled there as a result of their dissatisfaction with Japanese colonization, and remained there after World War II despite Japan's defeat. The postwar Soviet occupation of Sakhalin prevented any possible return by the Korean minority there, and others remained in Siberia, on the Manchurian border. Many Koreans who were living in Japan at the end of World War II settled there, although their legal status was ambiguous.[8]

But what can explain the fact that Koreans in Japan, as imperial subjects of the Japanese Empire, were suddenly re-categorized as foreigners at the end of World War II? Since modern nation states come into being as part of a global system, foreign citizens in those nation states are in principle not free to become citizens even if they wish to do so. Moreover, decisions surrounding citizenship are taken by the state, not by individuals.[9] As a result, Japanese imperialism legitimized national status in the sense that the term 'Japanese people' referred not only to the Japanese themselves but also included citizens from colonies who held Japanese nationality. This, of course, included the Koreans, due to the annexation of Korea from 1910 to 1945. But who was in charge of drawing the lines between the Japanese and their former imperial subjects, the Koreans, in the immediate postwar period? And what impact did the postwar emergence of nation states in place of empires and the re-definition of borders have on peripatetic former colonial subjects? What exactly was the colonial legacy that remained after the defeat of the Japanese Empire?

To answer these questions, we need to investigate Japanese colonization policies. In historical terms, the laws and attitudes towards border crossing have been shaped by the shifting political relations with Japan's neighbouring countries including Korea, China and Russia, and crucially by its relationship with the United States during the postwar decades.[10] Ironically, Japan's emergence as a colonial power in the 1890s was driven by the forced humiliation brought about by the inequalities in its treaty with the United States, and brought to a close by its defeat at the hands of that same Western power. This chapter traces the history of Japan's empire and looks at the way Japan conducted its colonization processes. Here, the shifts in Japanese colonization policies are analysed by reflecting on the changes in international relations which drove not only Japanese and Korean attitudes

The Annexation of Korea

The arrival of an American fleet at the Uraga Channel in 1853 under the command of Commodore Matthew Calbraith Perry provided Japan with a realistic view of both the international community and Japan's comparatively minor place in it.[11] But within twenty years of negotiating an end to its unequal treaties with the West, Japan had become a substantial colonial power. The revolutionary transformation of Japan from a weak, feudal and agrarian country into a modern industrial power that was economically and militarily capable of resisting foreign domination came about as a result of a collective recognition of its earlier vulnerability. To maximize Japan's strength, the effort to assert its presence in Asia and the consequent creation of empire would have to begin with the domination of neighbouring states, essentially those on the northeast Asian continent.[12] Most Asians argue that Japan's pursuit of its own security and economic survival was motivated by self-interest and achieved at whatever cost to broader Asian values. Most Japanese believe that the East could not have existed without Japan, and unless Japan was rendered invincible, there would be no way of saving Asia from the West. But making Japan invincible involved the economic exploitation of its neighbours.[13]

Japan's crucial justification is that the circumscribed location and dimensions of its empire were set by an overriding concern for Japan's insular security.[14] Eric Hobsbawm points out that, during the period from 1896 to 1915, around one-quarter of the world's land surface was distributed or redistributed as colonies among half a dozen states. During that period, Britain increased its territories by about 4 million square miles, France by some 3.5 million, Germany acquired more than 1 million and Belgium and Italy just under 1 million each. The USA, in comparison, acquired just 100,000.[15] Only Japan, China and Thailand were not colonized at that time, and after nearly a century and half of self-imposed isolation, few Japanese held any realistic views about either the international community or Japan's place in it. The arrival of an American flotilla in the middle of the nineteenth century, and Japan's eventual granting of unequal extraterritoriality to them, visibly demonstrated Japan's vulnerability to foreign aggression and stirred a debate over its national security.[16] Both Japan and China struggled to redefine their places in the new international order after they were forced to open up their long-isolated countries to foreign pressure. The first step on the colonization path for Japan was driven by the urgent need to consolidate and control a buffer territory to its peripheral regions.[17]

Major General Klemens Meckel, the Prussian advisor to the Meiji army, pointed out that from the perspective of the Meiji leadership Korea was viewed as a dagger thrust at the heart of Japan, and represented both a problem and an opportunity. As the Japanese government gathered strength throughout the 1880s, an array of diplomats, garrison commanders, traders and adventurers represented an increasingly aggressive Japanese presence in Korea, working to undermine the influence of China and the authority of the stubbornly traditional Korean government. At the end of the war with China in 1895, Ito Hirobumi and his co-negotiators acquired cession of the Liaotung peninsula and the island of Taiwan as well as the payment of a huge indemnity to Japan under Article 2 of the terms of the Treaty of Shimonoseki. However, the so-called Triple Intervention of Russia, Germany and France forced Japan to give Liaotung back to China within a few short years, and Russia later obtained from China a twenty-five year lease over the peninsula. However, Japan gained supremacy over Russia in 1905 and the lease was transferred to Japan with the consent of the Chinese government under the terms of the Treaty of Portsmouth (Article 5 and 6). Indeed, Japan's victory over Russia successfully countered an expansion of Russian influence southward from Manchuria into Korea. Japan offered St. Petersburg a compromise: recognition of Russian primacy in Manchuria in exchange for a free hand for Japan in Korea. The Portsmouth Treaty also gave Japan its next imperial acquisition, the southern half of Sakhalin – to which the Japanese gave the name Karafuto – and all rights and privileges in South Manchuria. This included the right to control the Russian-built South Manchuria Railway and lands adjacent to both sides of it.

These acquisitions increased Japanese pressure and influence on Korea,[18] and in consequence the Korean government had no alternative but to agree in November 1905 to the establishment of a Japanese protectorate to be administered by a *Tokan* (Resident-General). After Hirobumi Ito arrived in Korea in that role, his first move was a series of reforms which resulted in the systematic liquidation of Korean political institutions and their substitution with Japanese ones. In Ito's view, the most immediate concerns were institutional reforms addressed towards the elimination of official corruption and poverty in Korea.[19] Although he avoided discussing the possible annexation of Korea and is thought to have opposed it until his death in 1909, there were others within the government who regarded annexation as the most appropriate answer to Japan's security needs. For example, in February 1907, Ichiro Motono, the Japanese Minister in St Petersburg, wrote:

> We have to gradually move toward the goal of annexing Korea since there is no other way of insuring the establishment of tranquillity in Korea.[20]

Although Ito had presided over the dissolution of the Korean army and had forced the abdication of Emperor Kojong, his cautious policies were subjected to consistent criticism by the more bellicose members of the Japanese government.[21] The final years of Ito's residency were also characterized by increasing opposition from within Korea itself. In October of 1909, Ito was assassinated by a Korean named An Jung-geun while visiting the city of Harbin in Manchuria, an action which brought the threat of annexation much closer. In July 1910 the arrival of a new Resident-General in Seoul, army minister General Masatake Terauchi, signalled an end to even the pretence of Korean independence. The Treaty of Annexation was signed on August 22, and General Terauchi was subsequently appointed to the post of *Chosen Sotoku* (Governor-General of Korea), and all power suddenly became concentrated under his dominion over Korea and its people.[22]

In general, Western colonial powers in the nineteenth century pursued two contrasting alternatives when regulating their legal relations within their overseas territories. France adopted the *Système de Rattachement*, whereby countries such as Algeria were regarded as an integral part of metropolitan France, giving the African colony equal legal status to all the *departments* of the mother country. Britain, in contrast, viewed most of her overseas possessions in Asia and Africa in a more traditional colonial fashion, granting colonial governors broad powers to rule them.[23] When the Japanese acquired their first colony, Taiwan, in 1895, they were confronted with the question of which of the two systems of government they should use to determine their relations with the new territory. If Japan chose the British method of 'indirect rule', the colonial government of Taiwan would be free from all constitutional limitations, able to make its own decisions, enact its own laws and develop its own legal system. Taiwan would be a legally separate entity from metropolitan Japan, an idea clearly incompatible with the norm of 'one nation under one emperor.' The other alternative, the model used by the French in Algeria, would immediately bind all political and legal institutions of the colony to Japan, enforcing direct rule.[24] Edward Chen argues that naive confidence in their ability to transform the Taiwanese and Koreans into loyal subjects of the emperor led the Meiji leaders to reject the British system of colonial administration in favour of the direct rule approach. Takashi Hara, the Minister of Foreign Affairs in the colony rejected the British proposal and urged his government to regard Taiwan as an integral part of Japan in much the same way as Alsace and Lorraine were parts of Germany and Algeria a part of France.[25] But the extension of the constitution and the maximum application of Japanese laws were two vital steps necessary to achieve the goal of cultural and political integration. The outbreak of rebellion in Taiwan, its remoteness from Tokyo, and the difference in customs and traditions of the Taiwanese people were cited as reasons for requesting such an extraordinary measure.[26]

Mark Caprio argues that the equal inclusion of a colonized people as national citizens remains a vaguely defined goal, but strategic and malleable hurdles are usually imposed to ensure that all but a select few will always remain outsiders. In this regard, the colonizer develops rhetoric: on one hand, they preach the goals of internal colonization by seeking the dismantling of cultural barriers to aid the creation of a unified culture across a diverse array of peoples; on the other, they sustain a distinction between themselves and the colonized. Although Japanese assimilation policies required strict cultural assimilation, they failed to grant colonized citizens a political voice to pursue the same claims to liberties and economic opportunities as citizens of metropolitan Japan.[27] Marie Kim argues that, in Europe, the issue of assimilation or inclusion of natives was discussed as ideological and aspirational rhetoric, including the concept of individual enlightenment and self-improvement. In contrast, Kim maintains that, for the Japanese, the concept of assimilation did not aim towards the idea of holding the empire together. The colonized people would be forced to mirror the Japanese, not just figuratively but in every possible aspect; in terms of culture, spirit and loyalty.[28] Discrimination against colonial subjects could be justified not so much by their comparatively low levels of civilization than by their far lower levels of loyalty to the Japanese emperor. This is a key explanation for the particular exclusion of Koreans from political rights. Koreans were denied the same rights as the Japanese because the Japanese feared that ungrateful Koreans would undermine and eventually destroy colonial domination in the peripheries of the empire.

According to Chen, and as Montague Kirkwood suggested in 1895, if the Japanese government had allowed each colony its own legislative council, with power to enact its own laws and approve the budget of the colonial administration, the creation of a popularly elected legislative council in Taiwan would have attracted widespread support among the Taiwanese.[29] But instead of simply emulating the British style of indirect rule, Japan developed a complex legal compromise between the British and French systems regarding relationships between the centre and its such as Taiwan and Korea. As a result, following the French model of colonization, which insisted on homogeneity within an empire, Japanese colonial states faced a double constraint: the need to preserve Japanese nationalism under the Japanese Empire's 'sovereignty' and the simultaneous need to govern its indigenous subjects according to their own customs. This dichotomy had to be maintained until colonial subjects transformed themselves into diligent, loyal, law-abiding 'imperial peoples' (*komin*), imbued with the ethos of the Japanese state, bearing the same responsibilities and sharing the same life-styles as the Japanese. The dialectic of sovereignty and governmentality underlay the joint construction of the citizen and the native.[30] But how do we measure degrees of assimilation? The legal and cultural policies pursued by the Japanese towards the assimilation of non-Japanese Asians into their empire were always rooted in a profound ambivalence as to how far the assimilation process should go, and whether their colonial subjects could ever become true Japanese citizens.[31] Like the French in Algeria, Japan's overseas territories were occupied colonies where a minority of Japanese colonials existed amid a sea of indigenous peoples. Ultimately, the Japanese were not able to form a consistent theory of racial relations within their empire, and because of this they were unable to shape a coherent colonial doctrine which might justify the empire in their own eyes, to their subjects or to the rest of the world.[32] Thus the rights and legal status of individual Koreans remained historically ambiguous, a situation that simply by precedent was allowed to continue in the immediate aftermath of World War II.

Korea's Response to Japanese Colonization Policies

The first decade of Japanese colonial rule (1910–19) came to be known as the period of military rule (*budan seiji*) in Korea under the iron-fisted rule of General Masatake Terauchi (1910–16) and Yoshimichi Hasegawa (1916–19). Under the terms of an Imperial Rescript issued in 1910, Terauchi and his successors were invested with a wide range of discretionary powers which effectively made them the most powerful officials in the Japanese empire. Until 1919, the Governors-General of Korea were responsible to the Emperor alone, and even after that time were never fully subordinate to either the Japanese cabinet or the prime minister. Additionally, in order to exercise complete control over all civilian and military affairs, the Governor-General was authorized to issue *Seirei* (executive ordinances) which carried the same effect as laws passed by the Imperial Diet.[33] The colonial government under Terauchi developed into a powerful bureaucratic machine which was able to undertake a ruthless political, educational and social transformation of Korea whilst simultaneously attempting to obliterate Korean national identity. In the earlier stages of the colonial period, there were two important colonization policies; how to control Korea physically and psychologically; and how to obtain the highest potential benefits from agriculture and education.

During the initial phase of colonial rule, Japan attempted to develop Korea to its own economic advantage by seeking to dominate Korea's economic infrastructure. The objectives of Japan's economic policy in Korea were determined by the needs of the home economy; Korea was to be seen as a source of food (particularly rice) and raw materials, and as a market for Japanese manufactured goods. To achieve this, an economic environment consistent with these objectives had to be created. Before the Japanese protectorate, during the Yi period (1392–1910) Korea's economy had been dominated by agriculture. As late as 1910, nearly 85 per cent of all Korean households were engaged in farming.[34] In the absence of either a strong entrepreneurial class or a market economy, the primary source of wealth lay in the possession and acquisition of land. By the end of the Yi period, government control had been substantially eroded by the expansion of privately owned estates in the hands of *Yangban*[35] landlords. In order to reorganize the system of land ownership, the Governor General adopted Japanese civil law in 1912 and undertook an extensive land survey between 1910 and 1918 to obtain accurate data on all aspects of Korean agriculture. Japan's civil law provided legal grounds for the private ownership of land – including Japanese ownership – and the system was useful in helping to plan the development of agriculture. In addition to the identification, codification, and registration of all land, the purposes of the land survey were to determine and safeguard ownership rights, simplify the commercial transfer of land holdings, reform the tax system, provide the data necessary for planned agricultural development and rationalize landlord–tenant relations.[36]

Cumings, however, argues that these reforms did not seriously disturb the pre-existing pattern of land ownership.[37] Once the Japanese had established colonial rule, they found it advantageous to use Korean landlords to provide them with rice and keep the countryside stable.[38] Whilst the land survey was not harmful to the landlord class and other privileged sections in Korean society, it was peasantry who paid the price of Japanese colonial rule.[39] Much of Korea's rice land was owned by large-scale commercial owners, principally Japanese, who collected heavy land rents payable in kind. The changes in the land-tenure system provided structural alterations that supported the activities of large commercial corporations and provided the means by which agricultural production could be accelerated and exports to Japan increased. In 1910, only 17,000 tons of Korean rice were exported to Japan, but by 1919 this figure had increase to 400,000 tons, and by the mid-1930s the figure had reached nearly 1.5 million tons – more than 50 per cent of Korea's annual crop.[40] The principal victims of these forced exports were the poorer tenant families. The mass of the Korean peasantry found itself confronted with forces it could neither comprehend nor control. In debt to both landlords and moneylenders, who charged as much as 70 per cent annual interest, the peasant farmers' security of tenure was jeopardized and farmers were required to perform various types of unpaid labour. As a result, thousands of tenant-farmers abandoned their farms and sought employment elsewhere.[41] One of the most dramatic developments of this migration was the rapid emergence of a Korean immigrant community in Japan[42] (See Figure 2.1).

A further agenda was the erasure of Korean national identity. The keenest aspect of this agenda was the way Koreans were 'Japanized'. Throughout the colonial period, education was regarded as the principle instrument of assimilation. The ultimate objective of education, as enshrined in the Educational Ordinance for Korea issued in 1911, was the construction of loyal and good subjects in accordance with the Imperial Rescript on Education.[43] But the goal of Japanese educational policies in Korea was to duplicate the lower levels of the two-tier Meiji education system, the upper level being reserved for Japanese nationals residing in the colony. The reality was an education that was based on segregation between the Japanese and Koreans, with the latter being denied access to quality educational opportunities and post-secondary education. Although primary education was made compulsory for Japanese and Korean children alike, the Common School programme for Koreans was two years shorter than in Japanese Elementary Schools. A similar situation existed in the Higher Common Schools for Japanese pupils, while educational opportunities for Korean students beyond the secondary level were extremely limited.[44]

Japanese educational policies considered advanced training unsuitable for colonial subjects, for whom they primarily stressed teaching the Japanese language and offering vocational skills aimed towards the creation of semi-skilled employees in the agricultural, commercial and industrial sectors. Japanese language education became compulsory in all schools, while Korean language and literature classes were deliberately reduced or abolished. All schools in Korea required students to take ethics courses that taught the concept of loyalty to the Emperor and the state in order to turn Koreans into loyal subjects of the Japanese empire. Private sector education for Koreans, though not prohibited, was subjected to constant monitoring by the state. Under the Regulations for Professional Schools, issued in 1915, all private schools were required to obtain a government licence before they could offer education. Any school judged to be meddling in politics, encouraging anti-Japanese attitudes or refusing to use textbooks approved by the authorities was subject to immediate closure.[45] In the years that pre-dated Japanese colonization, Christian missionaries had established modern schools that emulated the principal features of schools in the West. Such schools were forcibly closed on the grounds that they were strongholds of actual or potential anti-Japanese sentiments. From the outset, governor general Masatake Terauchi made the Japanese position clear in a policy speech expressing Japanese fears regarding the connection between Korean private education and the national consciousness.

However, the acceptance of Japanese schooling did not lead to an unheeding loyalty to Japan. The situation for Koreans, who for years had endured discriminatory political structures and repressive colonial practices, became intolerable and eventually led to a mass uprising against the Japanese rulers on 1 March 1919. Known as the March First Uprising, or *Samil Undong*, the movement was a peaceful country-wide demonstration of the Korean desire for independence which quickly spread through the land and continued for months. The Japanese authorities responded brutally and violently, putting down the uprising with force. Japanese government statistics indicate 46,948 arrests, 7,509 killed and 15,961 injured.[46] But the March First Uprising not only exposed the limit of Japan's colonial influence but also showed in the long-term how Japanese policies of forced assimilation in Korea ironically produced the reverse effect, playing a central role in the formation of modern Korean nationalist consciousness as opposed to successfully eliminating Korean national identity.

In response to the uprising, Prime Minister Hara initiated a policy of conciliation known as Cultural Rule (*Bunka Seiji*). Under the 1922 rescript, all post-elementary educational facilities for both Japanese and Korean ethnicities were combined into one system, placing Korean ordinary and higher common schools on the same level as Japanese teaching establishments. The new Governor-General, Admiral Makoto Saito, also dismantled the hated gendarmerie system and replaced it with an exclusively civilian police force whilst simultaneously revising the regulations governing the appointment and promotion of Korean officials in order to expand employment opportunities for qualified Koreans. Meanwhile, all officials were advised to become more responsive to local needs and conditions.[47] Indeed, it is worth noting that in the aftermath of the March First Uprising, the enrolment rates in the common schools climbed sharply from 17.7 per cent in urban areas and 2.6 per cent in rural areas in 1915 to 33.8 per cent and 16.2 per cent respectively by 1926. By 1933 there were 680,000 students enrolled in 2,271 elementary schools and 35,000 students attending 579 rudimentary village schools in rural areas. Vocational education also expanded from 21 schools with 1872 students in 1919 to 52 schools with 9220 students in 1935.[48] These statistics show that many Koreans were accepting Japanese schooling in order to acquire the material advantages afforded by such education. The schools became a conduit through which Koreans climbed up the social ladder regardless of their class origin. The modern schools effectively dismantled the established Korean class system, opening up the possibility for children from lower social strata to gain access to occupations formerly denied them through education.[49] However, although the new curriculum effectively blurred the ethnic boundaries between colonizer and colonized, the hidden agenda behind the new approach to colonial education was aimed at teaching a form of 'learned passivity', the reward for which was measured in terms of the degree to which one accepted and lived according to the Imperial discourse. The colonial education system was thus based on the continuing efforts of the Japanese to train, educate and in the process bureaucratically degrade the colonial subjects to the level of de-politicized followers.[50]

Imperialism under the Slogan 'Japan and Korean as One Body'

The Japanese military takeover in Manchuria, which took place between 18 September 1931 and 9 March 1932, was a critical turning point in East Asian history. It marked the first surge of Japanese imperialism beyond the boundaries of its older colonial empire, and represented a process that had been granted tacit approval from the Western powers. It set Japan on a collision course with China, leading eventually to the Sino–Japanese War in 1937. The conflict broke down the old imperial status quo and led to a sudden and unprecedented expansion of Japanese power in Asia, which eventually led the nations of Southeast Asia into confrontation with the Western powers. Peter Duus, however, argues that any explanation of Japan's wartime expansion must take into account the international context. From 1931 onwards, Japanese military and political leaders projected

Japanese domination beyond the old colonial borders (Taiwan, Korea, the leasehold in Manchuria and the treaty port enclaves in China) by taking advantage of the continuing turmoil within China and the collapse of European colonial regimes in Southeast Asia.[51] The absence of any strong authority in China offered Japan the opportunity to seize control of the three north-eastern provinces in 1931–2 and encouraged further Japanese incursions into north China in the late 1930s. The outbreak of war in Europe and the early victories of Nazi Germany encouraged further Japanese aggression. When Nazi forces overthrew metropolitan regimes in the Netherlands and France, colonial governments on the periphery were cut adrift, with little guidance from the former centres and few resources with which to defend themselves.[52] By taking advantage of these opportunities, the Japanese managed to acquire formal or informal dominion over 340–350 million people by the end of 1942, covering a vast area stretching from the Solomon Islands in the mid-Pacific to Burma's border with India, and from the rain forests of New Guinea to the city shores of Attu and Kiska.[53]

What was going on outside Japan was always reflected by what was going on within its empire. External geopolitical events ultimately drove Japan's policy shift as far as the roles of its colonies were concerned. The early colonial regimes in Korea saw little point in building up a sophisticated industrial structure on the peninsula which would accomplish little more than duplicating home-grown structures. To Japanese colonizers in the early period, the primary economic function of Korea was to serve as an inexpensive export market for Japanese consumption.[54] But with the unfolding of Japan's war machine, the great mission required that both Korea and Manchuria should become a combined logistical and industrial depot that would supply whatever was required for the Japanese military effort.[55] At the same time, it became necessary for Korea to produce manpower to supplement that of Japan. Japan's 1935 population of 69 million was inadequate, and the Japanese hoped the Korea could be assimilated as easily as the Okinawans had been several decades earlier.[56] Japanese policies in Korea in the late 1930s began to show a distinct shift toward assimilation, and the demand for the unification of Korean and Japanese laws acquired new vitality.

THE ZAINICHI DIASPORA

With the outbreak of the Sino–Japanese War in 1937, the Japanese authorities in Korea used the slogan 'Japan and Korea as One Body' (*naisen ittai*) under which they accelerated and expanded their assimilation policy.[57] Under the slogan, the Koreans were expected to surrender their will totally to the Emperor and serve in his name;[58] this was referred to as the *kominka* policy. From 1937 to 1945, the *kominka* movement had touched almost every aspect of life in the colonies. Theoretically, whatever was deemed Japanese was to be imposed upon the colonial peoples, while whatever was considered Korean or Taiwanese was to be expunged in both colonies.[59] In other words, *kominka* can be seen as an extreme form of assimilation, little more than 'Japanization,' in deed and in spirit. Although Koreans were legally Japanese subjects, and although many of them had been provided with a Japanese education, the first act of Japanization was the eradication of any factors that distinguished the Koreans from Japanese. In his statement proclaiming Imperial Order Number 103 of the New Korean Education Ordinance on 4 March 1938, Governor-General Jiro Minami (who was in office from 1936 to 1942) announced the abolition of distinction between those who regularly use the national language and those who do not. Korean colonial subjects were thus forced to accept Japanese as the national language. Japanese language programmes were expanded and strengthened, while the limited number of Korean language courses that had been offered in the schools were dropped altogether from the school curriculum by 1938.[60] In addition, General Minami demanded that Koreans worship at Shinto Shrines, since these shrines were dedicated to the divine founder and ancestor of the emperor. Since 1936, not only were schoolchildren and college students taken to Shinto shrines for their regular monthly worship and on other occasions, but adults were also instructed to observe the bi-monthly Patriotic Days by marching to the shrines in throngs.[61] The next policy was the Name Change Order issued in February 1940, by which Koreans were forced to take Japanese names within six months of the proclamation. Those Koreans who retained their original Korean names were prevented from enrolling at school, refused service at government offices and excluded from the lists for food rations and other supplies.[62]

After 1937, the Japanese began to forcibly mobilize Koreans to work in support of the war effort by promulgating a General Mobilization Law and subsequently a National Conscription Ordinance in 1939. Koreans were initially drafted into *choyo* (labour draft) service, which was essentially non-combat labour that involved assembly line work and mining. These policies mainly attracted Korean peasants who left their rural roots and moved to the cities to work in industry. These poor, uneducated and unskilled Koreans were mobilized in Japan, Manchuko and northern Korea, and Irene Taeber estimates that by 1945 as much as 11.6 per cent of the Korean population was living outside Korea, most of them in Japan and Manchuko.[63] By 1945, hundreds of thousands of people – perhaps as many as two million or more – had left the land. They had uprooted themselves from Korea by abandoning their ancestral homes, and most of them had nowhere to go. It was those people who provided the foundation for the Korean diaspora; people who had been dispersed to places such as Japan, China, Sakhalin or Manchuko, where they found themselves settling down and remaining at the end of war.

However, as the war went on and resulted in mounting causalities, the Japanese sought more men to fight. In May 1942, the Japanese implemented a colonial conscription system called *chohei* (military draft) and drafted about 200,000 young Korean men to fight alongside the Japanese. By the end of World War II there were 186,680 Koreans in the Japanese army and 22,299 in the navy.[64] Chou argues that, besides coercion, the psychological factors used by the Japanese to promote young men's enthusiasm included the depiction of military service as the highest honour possible to bestow upon colonial subjects.[65] Park Chung Hee, former president of South Korea, was a typical example. One famous story tells of how Park demonstrated his loyalty by slitting his finger and writing a letter in his own blood directly to the Manchurian Military Academy pledging to die for the sake of the country. In fact, Park became a favourite of one of the Japanese army officers who supervised the school's military courses and was frequently held up as a model for other students.[66]

How did Japan manage to successfully mobilize Koreans not only physically but also psychologically? Japan's engagement in total war instigated a process of social change in Korea. As Japan's military, economic and political commitments grew more extensive, a labour shortage developed throughout Korea, offering more opportunities for peasants, workers and educated Koreans than ever before. Before the 1930s the colonial authorities had shown little interest in developing Korea's technological base beyond a rudimentary level. In the aggressive and warlike atmosphere of the mid-1930s, however, the Governor-General launched a programme to upgrade the colony's vocational and technical education at all levels, from elementary schools to Keijo Imperial University. Since the Japanese were in great need of assistants who spoke their language and who were at least more trustworthy than the natives, most Koreans naturally seized whatever opportunities for advancement became available to them under the colonial system.[67] Koreans who were not drafted until very late in the war benefited from the shortage of Japanese manpower in white-collar positions. For example, the number of Koreans of high rank holding key official posts increased by 400 per cent, while those of junior rank grew by about 150 per cent, with the most dramatic increases coming after the establishment of Manchuko and later during the war in China.[68]

As a result, the *kominka* movement, under the slogan of *naisen ittai* ('Japan and Korean as One Body'), created a dichotomy of feeling between colonial resistance as outsiders and internal trends towards empowerment, which continued to expand until the abrupt end of World War II. Cumings explains that the colonial period brought forth an entirely new set of Korean political leaders, spawned both by the resistance to and by the opportunities of Japanese colonialism.[69] Cumings suggests that Koreans provided Japan with a stark contrast to its other colony in Taiwan. Chou argues that the *kominka* movement may have succeeded in making them less Chinese while failing in converting Taiwanese into true Japanese.[70] There had always been a strong movement for independence in Korea, but independence in Taiwan, if it had ever been considered at all, had been evidently regarded as hopeless, not even worth thinking about.[71] Korean nationalists, on the other hand, were split between those who remained in Korea and those who went into exile abroad. At home, the nationalists were divided into radicals and gradualists. The latter urged a path of preparing Koreans for independence through cultural and educational activities.[72] Even the exiles were split between people who favoured a militant, armed struggle and those who urged diplomatic methods for securing Korean independence.[73] The first President of South Korea belonged to the latter group, whilst the founder of North Korea, Kim Il Sung, emerged from the former. So the *kominka* movement not only failed to turn Koreans into true Japanese, but also contributed in the end to creating two divided nations which have lasted until the present.

The Legacy of Japanese Colonization

Colonial legacies can be regarded as features endowed by influential colonial rule, such as patriotism on a country-wide scale and a sense of belonging to a wider empire with a shared value system. Whether such influences remain in post-empire periods depends on the length of the influence and its strength of its bond. Mark Peattie argues that in Korea, colonialism lasted just long enough for Koreans to recall the national injury suffered at the hands of the Japanese but not long enough for the effects of Japanese education and assimilation policies to influence the loyalties and interests of a younger generation of Koreans.[74] Although Koreans lost their homeland for a comparatively short period, its recovery during the post-World War II period proved to be an extremely complicated process.[75] Even though the length of colonial rule may not have been sufficient to destroy Korea's individuality and sense of national pride, the strict bond of the forced colonial assimilation policies was so intense that the sudden vacuum it left when Korea's colonizer was defeated left a great impact on Korean society.

The tragedy for Korea was that its independence was achieved only by the defeat of Japan in World War II, rather than at the hands of the Koreans themselves, and it was widely held that the Koreans were not ready for independence when the war ended so abruptly. One of the reasons for that is the memories of the final years of the Yi dynasty did not inspire confidence in the Korean potential for effective self-government. According to US Navy Department analysis in 1943, it was considered that the country was bereft of leadership, and that Korea could fall into a political vacuum often seen in post-empire colonies in the early stages of self-government.[76]

Into that vacuum stepped two politicians of different views and from different backgrounds. Syngman Rhee, the first president of South Korea, had previously acted as president of the provisional government based in Shanghai in 1919, although the bureaucrats of the Roosevelt administration regarded him as a tiresome person of no real importance. Later, he became acquainted with Preston Goodfellow, who worked at the time for the Office of Strategic Services (OSS). Goodfellow played an important role in advancing Rhee's career in the immediate aftermath of World War II, but the official American view was reflected in a report compiled for the American Navy Department which stated that Rhee and his Korean colleagues were frustrated and unemployed men.[77]

Kim Il Sung, on the other hand, was not to emerge as a major personality until after 1945. He had previous experience in his formative years of cooperating with the Chinese and the Russians, but he had no experience of working with communists in Korea. He moved from Manchuria to Khabarovsk in January 1941, then probably fought with the Soviet forces, although there is no evidence to substantiate the grandiose claims Kim was later to make concerning his contributions to World War II.[78]

Given these two leaders, one considered a frustrated southern leader kept in check by American dominance and the other a charismatic northern communist ally, the seeds were sown for Korea's future. Lee Won-sul argues that even before the arrival of the American occupation forces Korean society had been polarized, and this political polarization during the short interregnum period laid the foundations for a chaotic and troubled future.[79] Cumings points out that colonial rule left Korea with no single indigenous leader, nor any leadership legitimized by popular support. The crucial weakness of Korean communism in the 1920s and 1930s was caused by inadequate leadership, jealousies and animosities. From this perspective, those Koreans who lived in exile and sought to dominate the political stage in postwar Korea had personal backgrounds and political histories that were at best obscure.

There are various academic debates on the origins of the Korean War. On one level the war came about because of the domestic struggle between two different political viewpoints and a clash with in the Korean social consciousness.[80] On another level, Korea's geographical position and political polarization was seen as a delicate one by the Russian and American superpowers who were themselves rapidly polarizing in preparation for the Cold War. The superpowers quite possibly overestimated the strength and impact of strikes, rebellions and even guerrilla warfare in Korea, eventually concluding that international intervention was the only possible solution to Korea's domestic problem. Some scholars argue that the Soviet Union initiated the Cold War with offensive aggression in their search for a communist world revolution, and the United States acted accordingly to block Soviet expansion. Thus the Korean War was the inevitable outcome of an international power struggle, in which the United States saw itself as an international policeman, blocking the dominance of Soviet influence in the fragile Korean peninsula and striving to contain further expansion of Communism elsewhere.[81] Cumings argues that whether we speak for the Korean nation or for class structure more generally, the great shifting and reintegration occasioned by the Japanese defeat had critical consequences, which must be seen as the touchstone for evaluating the politics of a liberated Korea and the origins of the Korean War.[82] Peter Lowe considers that there are essentially two viewpoints on the political and social evolution of Korea between 1945 and 1950 and Korea's place within the emerging Cold War.[83]

How much of the Korean War was civil and how much of it was international? Either way, Japan's colonial legacy and its abrupt demise was an indispensable key factor for the subsequent internal division of Korea. The internal chaos in Korea created by the unsuccessful liquidation of Japan's colonial legacy and international involvement during the first few months of the occupation period proved a formidable barrier to the Korean process of nation-building.[84] The political leaders were too weak in their preparation and too dependent upon the rapidly polarizing superpowers to overcome the obstacles preventing them from building an independent nation, and the internal issue of selecting a form of unified government therefore became internationalized. From this concept, Masao Okonogi defined the Korean War by focusing on the links between domestic and international political factors:

> The Korean War was a complex struggle in which domestic and international elements were intertwined. It was the consequence of historical processes both inside and outside Korea that emerged in the summer of 1945 and the Korean War arose out of a vicious circle of escalation between domestic and international forces, whereby domestic politics were 'internationalized' and international politics 'internalized.'[85]

From Japan's point of view, the failure of their colonial assimilation policy is proved in terms of colonial legacy by the sizable Korean diaspora that remained in Japan at the end of World War II. By the time of the Japanese surrender, there were approximately 2.2 million Koreans in Japan[86] and more than 1,548,000 Koreans in Manchuria. As stated above, that the principle causes for the Korean migration to Japan were twofold; firstly, the colonial policies that led to the rapid growth of a large impoverished and landless peasant class in southern Korea, and secondly the considerable proportion of the excess rural labour that was those policies created. As Japan's military expansion advanced, the labour market in Japan was unable to meet the demands of industry and compensated for labour shortages by recruiting relatively cheap labour from Korea. Although the US military governments in Japan and Korea facilitated the repatriation of nearly one-and-a-half-million Koreans between 1945 and 1949, some of them, 600,000 decided to remain in Japan. Most of them had uprooted themselves from Korea by abandoning their ancestral homes and had nowhere to go and nothing to return to. In stark contrast to the colonization period, the framework of law and institutions in Korea was no longer determined by the Japanese government. Instead it was designed through close collaboration between the allied occupation authorities and the Japanese state. While the allied occupation authorities in Japan were eagerly encouraging Koreans to go home, the US Military Government in Korea (USMGK) was doing little to prepare for their arrival. As a result, the dispossessed Koreans considered it preferable to stay where they were rather than face an uncertain future in Korea. Although the Japanization of Korean youth required converting the Korean consciousness into a Japanese one, a speech by Japanese Diet member Saburo Shikuma on 17 August 1946 confirmed the argument that the colonization period had been too short for the effects of Japanese education and assimilation to have taken root within the younger generation of Koreans:

> We refuse to stand by in silence watching Formosans and Koreans who have resided in Japan as Japanese up to the time of surrender, swaggering about as if they were nationals of victorious nations. It is most deplorable that those who lived under our law and order until the last moment of the surrender should suddenly alter their attitude to act like conquerors, posting on railway carriages 'Reserved' without any authorization, insulting and oppressing Japanese passengers and otherwise

committing unspeakable violence elsewhere. The actions of these Koreans and Formosans make the blood in our veins, in our misery of defeat, boil.[87]

Saburo's rhetoric strongly suggests that the sizable Korean minority had become a key agenda that the Japanese government was compelled to deal with during a time in which it was striving to recover its own sovereignty whilst eliminating its own pre-war colonial legacy.

3

No Alliance and a Strong Historical Legacy: Exclusionary Policies towards the Zainichi in the Post-World War II Era (1945–64)

Introduction

In the opening sentences of his book *Nations in Alliance*, George Liska argues that 'it is impossible to speak of international relations without referring to alliances.'[1] But what interests can nations pursue through alliances? While some nations are content to preserve the status quo as long as they remain satisfied with their continuing development, others may not be satisfied for a variety of reasons. Before they form any alliances, nations may possess similar levels of autonomy but varying levels of security. Whilst all nations strive to possess similar degrees of autonomy, minor powers will concede some degree of autonomy if they choose to form an alliance with a major power in return for the security that major power provides.[2] This is categorized as an asymmetric alliance. The most extreme form of asymmetrical alliance is a security guarantee, by which a major power agrees to protect a smaller state but does not require or expect that smaller power to do much in return.[3] In the early stages of the postwar era, the alliances between the US and both Korea and Japan were examples of this extreme form of alliance.

After World War II, the highest priority for states such as Japan and South Korea was their security. Both countries became strong allies of the

United States, and as they looked to America for security and protection their relationship with each other was of secondary concern and remained unstable, with neither country seeking to normalize its relations with the other. From the US viewpoint, this long-term instability was preserved and utilized – and perhaps even encouraged – as part of a hub-and-spoke alliance structure (a term coined by John Foster Dulles) as a hedge against what the US viewed as an undesirable multilateral order that might emerge in the region if Japan and South Korea were to normalize their relationship.[4] To enable major powers to retain control over less powerful members of an asymmetric alliance, bilateral control is a more direct and effective means of exercising power.[5] East Asia's security bilateralism today is therefore a historical artefact – a remnant of postwar American rationale for constructing alliance networks in Asia.[6]

Historically, the United States had shown little interest in the Korean peninsula. Although Woodrow Wilson's advocacy for self-determination had become well known among educated Koreans before 1905, the United States had been reluctant to intervene and had expressed no interest in supporting Korean independence, emphasizing its own neutrality and its policy of non-interference.[7] Many US officials, including Wilson himself, had supported the common idea that Korean problems, including those relating to China, were a low priority compared to more pressing issues in Europe.[8] Franklin D. Roosevelt was another American politician who gave little thought to Korea during those years. Towards the end of World War II, he proposed placing the whole peninsula under the temporary trusteeship of China, the Soviet Union, Britain and the United States, and relied on the pledges of Chiang, Stalin and Churchill that Korea would in due course be granted independence.

What changed the long-held perceptions of Korea as relatively unimportant, turning it and Japan into important allies in the larger context of American regional and global policy? The origin of the 38th parallel as Korea's dividing line has never been subject to academic debate. The widely held view is that the dividing line originated after World War II in a proposal by the United States War Department, with the aim of facilitating the surrender of Japanese troops whilst preventing the Soviet occupation of the entire peninsula.[9] In other words, the 38th parallel was intended to mark a temporary line of demarcation, dividing the responsibility for

NO ALLIANCE AND A STRONG HISTORICAL LEGACY

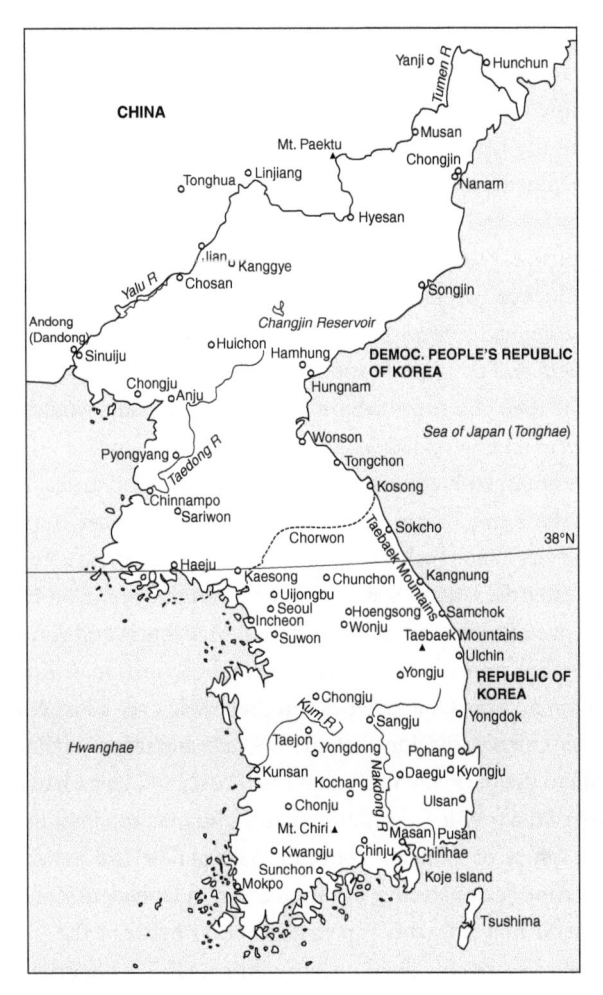

Map 3.1 The 38th parallel[10]

carrying out the Japanese surrender between the United States and Soviet Russia (Map 3.1). As US President Harry S. Truman recalled, there was no thought of a permanent division.

While the proposal to divide the peninsula for administrative purposes was created by the United States in talks with Soviet Russia, the 38th parallel had been historically established as a dividing line in Korea long before the end of World War II. During the high tide of Russian influence on

Korea, Japan proposed in 1896 that Korea be divided along the 38th parallel. Russia rejected this division at the time because it still hoped to gain control of the entire country,[11] but in 1898 the Russian minister to Korea proposed to the Japanese minister that the two powers should divide this strategic peninsula along the 38th parallel, with Russia taking the north and Pyongyang, and Japan the south with Seoul. The proposal eventually fell through when war broke out between the two nations (1904–5), and Russian influence was ultimately ended by defeat.[12] While military planners in the Pentagon were unaware of these precedents, Stalin was well aware of them and their implications on the 38th parallel, and he was satisfied with the idea that the northern half of the Peninsula would become a buffer zone, as it safeguarded Soviet control over Manchuria.[13] That was the reason why Stalin willingly accepted the division of Korea, despite the fact that the Soviet Army was in a position to occupy the entire country at the time. The Soviet Union had entered the war against Japan on 9 August 1945, three days after the United States dropped an atomic bomb on Hiroshima, and had moved Russian military forces into Manchuria and Korea.[14]

From Stalin's perspective, the division and occupation of Korea by the Soviet Union and the United States was preferable over a longer timescale than the temporary division that the US had envisaged. Stalin hoped it would lead to the permanent division that Russia and Japan had discussed at the turn of the twentieth century. As such, Russia's apparent intention to establish a sphere of influence above the 38th parallel did not accord with the earlier American objective of a united and independent Korea.[15]

When and how was this gap in perception between the US and the Soviet Union towards the dividing line eliminated? It is important to know this, as the point of elimination marks the turning point at which the underlying dynamics of geopolitical conflict began to emerge; in other words, the establishment of the 38th parallel as a dividing line in Korea is a pivotal point of conflict between the Great Powers, who then sought to preserve the dividing line as a symbolic buffer that separated the two Koreas – even in the face of Korea's own goal of national unification. Although the reason for sending troops into Korea was initially to disarm the Japanese forces there and maintain order below the 38th parallel, the United States quickly found itself in the position of being the principal sponsor of the ROK, established in 1948. As the Cold War began, America

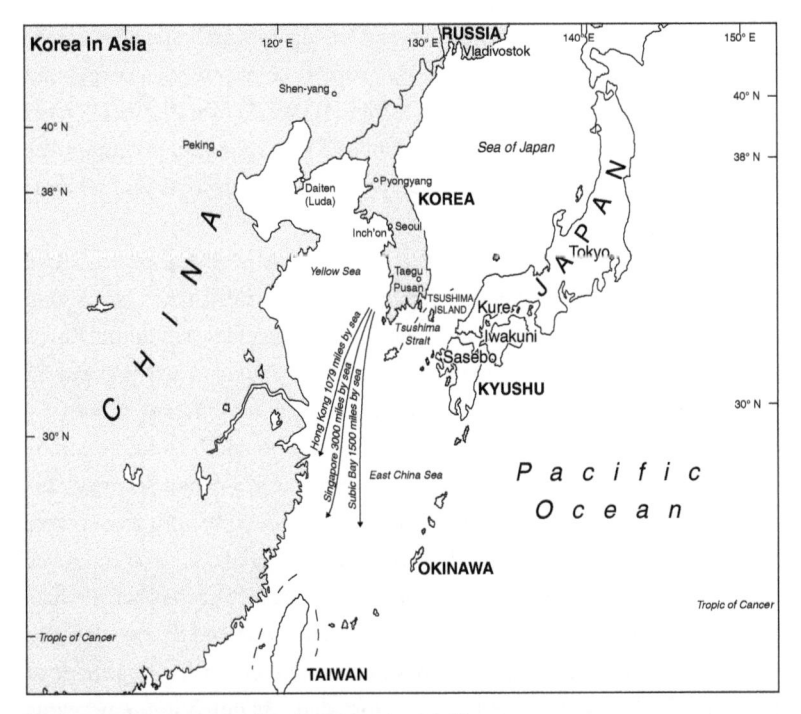

Map 3.2 Korea's geographical position in Asia[16]

started to act as Korea's protector from external military threats, and as the provider of assistance for its economic sustenance and eventual growth.[17]

While the US and the Soviet Union staked their claims as soon as the Pacific War ended, Japanese leaders withheld almost all judgement on the political situation in Korea. The Korean War would serve to stimulate rearmament and economic reconstruction in Japan a few years later, but for most Japanese it was merely 'distant fire' which aroused no greater awareness of Korea.[18] From the Korean standpoint, foreign interest was shifted from the previous protector, Japan, to the current saviour, the US, which recognized Korea's broader significance in terms of strategic planning. Interestingly, the geopolitical conceptions that governed the behaviour of the Great Powers showed remarkable continuity with conceptions of the past: that Korea represented a vital borderland – not only between Russia, China and Japan as before, but now also for the United States in its global role and its growing stance against communism.[19] The perception that the

Koreans were too weak to protect themselves against their stronger neighbours had been used as Japan's justification for colonization, as explained in the previous chapter. Now, after the end of World War II, the US used the same reasoning, and replaced Japan as Korea's mentor, taking on the vital task of preventing the peninsula from falling into unfriendly hands at whatever cost.

How did US intervention affect relationships between South Korea and other countries – especially Japan? And why did it take such a long time – almost two decades – to formalize relations between South Korea and Japan by ratifying the Japan–ROK treaty in 1965, a treaty that put an official end to the history of colonial domination and marked the a new start for diplomatic relations between the two countries? These questions are deeply interrelated with the formation of the US–Japan security alliance and US–ROK alliance treaty. Although Japan and South Korea were viewed in the same light under the US security umbrella, the two countries had no interest in forming normal diplomatic ties with each other, a reluctance that underlined the fact that they remained unallied for two decades after the end of World War II. This situation influenced the treatment of the Zainichi diaspora by the Japanese host state. As noted in the previous theory chapter, this situation tests if the host state and homeland have no diplomatic relations, then the host state policy toward the diaspora group perceived to be backed an enemy homeland will be *exclusion*. This chapter will therefore explore the conditions under which Japanese elites selected exclusionary policies toward the Zainichi diaspora in the postwar era from 1945 to 1965.

In order to identify how and why the US changed its perception towards Korea and Japan, we need to trace US foreign policy making process from the end of World War II and explore the formation of separate security alliances between the US and Japan and the US and the ROK. The US concern with and involvement in Korea changed during the period between World War II and the Korean War (1945–53), a period that ended with the US–ROK Mutual Defence Treaty. However, Japan failed to establish its own diplomatic relations with the ROK during the period between regaining its own independence in 1952 and the normalization treaty in1965. Japan formed its own security alliance with the US, an alliance that not only served Japan's geopolitical security but also established the

US as its principal Asian political ally and critically contributed to South Korea's military defence against the communist bloc. But despite this, Japan's continuing absence of bilateral relations with South Korea allowed Japanese policymakers to pursue exclusionary policies toward the Zainichi diaspora, despite the fact that they both shared a major power, the US, as a common allied country.

The Korean War and the US–ROK Alliance

The South Korean state would never have come into existence in 1948 without American intervention, nor would it have survived the hardships brought on by national division and the horrific war that followed without vast US military and economic assistance.[20] For the United States, building and stabilizing South Korea came at an enormous cost in terms of both material resources and human lives before the US could finally embrace South Korea as one of its client states in the Asia–Pacific region. By concluding a mutual defence treaty with Korea as an essential link in its regional containment system, the US continued to play a vital role in guaranteeing peace and stability on the Korean Peninsula.

Among the Great Powers bordering the Korean peninsula, such as Japan, Russia and China, Korea has traditionally been viewed as an east-west stepping-stone of considerable strategic value. As World War II approached its end, Japan's power on the peninsula and its desire to use Korea as a springboard into Asia diminished, and its influence became minimal. China was caught up in its own civil war until 1949, so it might be reasonable to assume that, as World War II ended, the Soviet Union was the sole Great Power bordering the Korean peninsula that was in a state of readiness to recover the assets in Asia that Japan had seized nearly forty years earlier. It could be argued that Soviet postwar planning for Korea began around the time that Stalin eventually made up his mind to enter the war against Japan in October 1943.[21] Indeed, Soviet leaders recognized the profound geopolitical significance of Korea and were acting consciously in accord with their pre-war dynamic to secure its realignment.

But to the United States Korea was a nation located thousands of miles across the Pacific. Not much bigger than the state of Minnesota, it remained chiefly of only regional importance. For US military leaders,

Korea was a rear-guard defence area, not one for frontal aggression or expansionism: Okinawa held a much more important place in US postwar strategy. William Whitney Stueck argues that policymakers in Washington during the Roosevelt era intended to achieve victory at the lowest possible cost to American lives and resources, and saw unilateral American occupation as an undesirable responsibility.[22] Initially, American leaders viewed cooperation with the Soviet Union as a prerequisite for achieving their goals, but Washington became preoccupied with the dangers of over-commitment to a country whose population was numerous and nationalistic, where the prospects for long-term exploitation by a foreign power were consequently poor.

That viewpoint, however, was a short-sighted one insofar as it gave no consideration to the possibility that the Soviets would strive to enhance their own power in Northeast Asia now that Japan, the traditional counterweight to Russia in the area, had been removed. This possibility became even more realistic after Roosevelt's extensive concessions to Stalin at the Yalta agreement, concessions that gave Russia the southern half of Sakhalin and the adjacent islands, the Kuril Islands as well as former Tsarist rights in the Chinese port of Dairen. Roosevelt also conceded the USSR's pre-eminent interests in Manchuria, especially the Chinese Eastern and South Manchurian railways, and also proposed at Yalta in early 1945 a multi-power trusteeship for Korea to include the United States, the Soviet Union, China, and Great Britain. At the November 1943 Cairo Conference which outlined the allied position against Japan, China and Great Britain agreed with the American initiative that Japan should be stripped of its colonies, and that in due course Korea should become free and independent. However, from the US perspective, Korea was thought too immature politically to manage the problems of immediate independence; it was considered instead that trusteeships could teach states that were unused to governing themselves the art of self-rule.[23]

Following the death of Franklin D. Roosevelt, whatever trust had existed in Soviet–American relations disappeared in the wake of disagreements over Eastern Europe. President Truman[24] devised a strategy that appeared to ensure the realization of US ambitions for Korea. Washington's objective had been the creation of an independent, united, western-oriented nation that would possess a progressive and democratic government. If American forces liberated Korea unilaterally, Truman reasoned, and then the United States could reconstruct this Asian nation without Soviet interference. By late July 1945, Truman hoped that the atomic bomb would quickly force Japan's surrender before the Soviet Union entered the war and thus avoid negotiations regarding the peninsula. But the Soviet Union entered the war on 9 August, well in advance of the anticipated date, and therefore it already had troops in Korea when Japan surrendered. American forces did not arrive until 8 September, and the United States suffered from the consequent lack of concrete agreements.[25]

At the Moscow Conference in December 1945, the United States and the Soviet Union appeared to agree on international trusteeship as the best method for resolving the Korean problem. The Soviets suggested establishing a joint commission to bring about economic unification, a Korean provisional government and a four-power trusteeship – the Soviet Union, the United States, Great Britain and China.[26] The Joint Commission convened on 20 March and rapidly became deadlocked on the question of consulting various Korean groups. The Russians insisted on consulting only groups that supported the Moscow agreements. The Americans objected because the acceptance of the Russian demand would eliminate from consultation everyone except the leftist groups under Soviet influence. The stalemate reflected unwillingness on either side to accept unification procedures that might result in an unfriendly Korea.[27] When Stalin refused to accept the American interpretation of the Moscow decision, Truman rejected further negotiations and ultimately turned to the policy of containment in an effort to break the deadlock. Indeed, the direct involvement of the superpowers in Korea and Russia's apparent determination to communize the North – and possibly the entire peninsula – led the Truman administration to recognize a substantial American stake in the country. A show of weakness there might undermine the credibility of the United States worldwide,[28] so the Truman administration felt compelled to view Korea as a buffer protecting US security in the pursuit of its regional and global strategy, and America therefore began to represent itself as a new power in Korea, in accordance with the pre-war geopolitical competing dynamic.

However, Truman's strategy for containing Soviet expansion in Korea was limited and required only that the United States provide economic aid, technical advice, and small amounts of military assistance.[29] The estimated cost of the three-year programme allocated to Korea in the proposed US War Department budget was $600 million, which was considered too expensive to put into effect.[30] There were many factors that blocked the proposed aid programme as it went through Congress. The first was a growing concern regarding the apparent inability of the Korean people to govern themselves. Korea was in a state of considerable turmoil during 1947 and 1948, a condition that was disturbing both State and War Department officials in Washington. When American forces occupied the southern half of the Korean peninsula in 1945, two distinct possibilities for the creation of institutions existed. One was aligning with the Korean left, the Korean People's Republic (KPR) party which had formed a provisional government with strong popular support after Japan surrendered to the Allies at the end of World War II.[31] The other was to reconstruct the vast centralized bureaucratic structure that the Japanese had used to govern Korea. American officials doubted the capacity of Koreans to govern themselves and suspected that the KPR was connected to international Communism. Thus, the US Military Government in Korea opted to govern through the old colonial power structure rather than back the KPR.

The Americans filled many of the highest bureaucratic posts with Koreans who had been affiliated with the KPR's rival organization, the Korean Democratic Party (KDP), a group of Korean conservatives, some of whom were tainted by collaboration with Japanese colonialism.[32] The occupation commander believed that, with Syngman Rhee's cooperation, a strong coalition of non-Communist political groups who were sympathetic to American policies might emerge. However, the US failed to construct such an institution and instead was forced to content themselves with reconstructing the pre-existing colonial bureaucracy. Economic conditions aggravated the unrest further. Cut off from traditional commercial ties with the north and Japan, Korea had difficulty sustaining its growing population. Railways deteriorated from a lack of spare parts, and the availability of electricity (the sources of which were largely in the north) declined. By January 1947, the military government's civilian supply programme had been halted because of a lack of funds, and the distribution of rice to the general populace was becoming more and more difficult.[33] Although strong repressive actions from August onwards had begun to weaken the Communist party, the social and economic problems that bred unrest remained unresolved.

A second area of opposition centred on the occupation cost, emerging from budget-conscious congressmen in Washington. Under-secretary of War Robert P. Patterson and senior assistant Howard C. Peterson were doubtful that Congress would approve the planned $600 million for Korea, explaining that such expenditure would be inadvisable in the light of more important threats such as a possible emergency in Europe and needs associated with that region.[34] The planners in Washington still remained fixated on a Europe-first strategy. Patterson agreed with Petersen's analysis; he did not openly oppose the Korean aid programme, although he used his reservations as an instrument to gain concessions from the State Department on other aspects of Korean policy such as early withdrawal from the peninsula.[35] Patterson drew the logical conclusion, stating: 'I am convinced that the United States should pursue forcefully a course of action whereby we get out of Korea at an early date, and believe all our measures should have early withdrawal as their overriding objective.'[36]

The desire for early withdrawal of US troops from Korea was driven by a declining interest in China since 1946, which ultimately led to the withdrawal decision. Adam Ulam pointed out that Washington's failure to intervene on a massive scale to resist the Communist advance in China made it difficult to deliver a credible image of American determination in Korea.[37] However, that could be overcome by maintaining a strong position in Japan. With the collapse of the Nationalist Chinese on the Asian mainland in 1949, Japan was gradually replacing China as the country the United States was hoping would become its principal Asian political ally.[38] Patterson also acknowledged that 'from the standpoint of US security, our policy in the Far East cannot be considered on a piecemeal basis, and logically the policy concerning Korea must be viewed as part of an integrated whole which includes Manchuria and China.' His argument was based on the logic that, given the decreasing capacity of the US military, 'the United States must review critically all programmes with the realization that non-availability of means will force us to drop the least beneficial of them in the near future.'[39]

Patterson's opinion supported the idea that the earlier American proposal for the 38th parallel was motivated only by the aim of preventing

Soviet occupation of the entire peninsula. There was no clear strategic blueprint as to how the difference between the 38th parallel and the 39th parallel implied any geopolitical advantage that the US might have gained in future. In fact, the 39th parallel would probably have been a better choice, enabling the American military forces to maintain the shortest possible boundary line.[40] However, Stalin was satisfied with the creation of a buffer zone in the northern half of the Korean Peninsula that would safeguard Soviet control over Manchuria and allow the Russians to control the northern half of Hokkaido (part of Japanese territory)[41] (Map 3.2). America's lack of decisive planning demonstrated the unpreparedness of US military planners (unlike those of the Soviet Union) as well as the absence of a comprehensive strategy towards East Asia from the earlier period of the occupation. This lack of cohesion influenced US policies in East Asia, ultimately contributing to the outbreak of Korean War. In accounting for the origins of the North Korean attack of June 1950, Nikita Khrushchev also suggested that more extensive symbolic acts by the United States might well have deterred Stalin from giving Kim Il Sung the go-ahead for an invasion of the south.[42] It is no wonder that the North and its Soviet – and presumably its Chinese backers too – perceived an extraordinary weakness in their opponents, making the risk of war acceptable.[43]

By late September 1947, therefore, a consensus had emerged among State and Defence planners in favour of a graceful withdrawal from Korea. If this was to be done without simply abandoning the peninsula to Communism, new initiatives had to be taken to end the Soviet–American deadlock and to establish an independent government in South Korea without a major American military presence. Any evaluation of American policy towards Korea in the fall of 1947 should therefore begin with the United Nations. The UN called for free elections throughout the Korean Peninsula under the supervision of the new nine-nation United Nations Temporary Commission on Korea (UNTCOK), but Stalin refused to allow the Commission entry into North Korea, dismissing it as a compliant tool of US policy and rejecting the concept of free elections in North Korea.[44] American leaders considered that the UN would be more likely to handle the problem effectively because it would receive wider publicity there. There was an advantage in launching an independent government for South Korea through the UN rather than unilaterally. Sponsorship of such a government by the largest organization of states in the world would add legitimacy to the project. The failure of the United States and the Soviet Union to reach agreement on unification looked to be on the cards, and because Washington regarded a simple American withdrawal as inadvisable, submitting the Korean problem to the United Nations prior to the exhaustion of other approaches appeared a more attractive proposition.[45]

The UNTCOK-supervised elections were held in only the US occupied zone on 10 May 1948, after which the ROK under President Syngman Rhee was proclaimed on 15 August 1948. President Truman appointed John J. Muccio, a career diplomat who had served in Shanghai, Panama and Germany, as the first American ambassador to South Korea. The United States also co-sponsored a resolution adopted by the United Nations General Assembly in Paris on 12 December 1948, which stated:

> There has been established a lawful government having effective control and jurisdiction over the part of Korea where the Temporary Commission was able to observe and consult and in which the great majority of the people of all Korea resided.[46]

The resolution accorded a degree of international legitimacy to the ROK, recommended the withdrawal of occupying forces from the country as soon as possible and reorganized UNTCOK into a seven-nation United Nations Commission on Korea (UNCOK), whose primary responsibility was to promote the peaceful unification of Korea.[47] In the Soviet-controlled zone, the DPRK, under Premier Kim Il Sung was also established on 9 September 1948. As a result of those actions, the 38th parallel became completely closed and was effectively turned into a frontier.[48] The Truman administration completed the withdrawal of US troops from South Korea by the end of June 1949, with the exception of the 495-man Korean Military Advisory Group (KMAG).[49] The withdrawal was ultimately made despite pleas from President Rhee, in a letter to President Truman on 19 November 1948, stating that it was imperative for the United States to maintain its occupation forces for the time being and to establish a military and naval mission as a deterrent to aggression and consequent civil war.[50]

The Korean War, however, dramatically reversed US security policies in Asia. Ultimately, US perceptions of Communist Chinese strength finally began to influence policy makers in Washington. Washington had viewed China as a difficult area to influence, especially in light of its proximity to Korea. China's lack of industrial development and its limited natural resources had kept it well down the list of priorities for American attention, and State Department observers had regarded the country under nationalist leader Chiang Kai-shek as being too weak economically and militarily to sustain a victorious military offensive against the Communists, even with large-scale American aid. Although the Truman administration never cut off aid to the Chinese National government, the gradual withdrawal of American troops from China during 1946 was aimed towards limiting American commitments there. In fact, the United States strenuously avoided large-scale intervention in China, even when the Communists marched to victory between 1947 and 1949.[51] The massive Chinese counteroffensive in Korea in late November 1950 allowed the US to admit its own miscalculation that neither the Soviet Union nor Communist China would intervene directly in the conflict. China unexpectedly entered the war, although the Soviet navy never interfered with US naval operations.

The most significant effect on international relations caused by the Korean War – especially regarding the United States and its relationships with the USSR, the People's Republic of China (PRC) and the DPRK – was to introduce China as a major actor on the Korean scene.[52] In terms of former relations, the Chinese entry into the war led to a series of US countermeasures, which ranged from trade embargoes through the network of US bases and alliances constructed in order to contain China to a sustained US effort to isolate the PRC diplomatically. At first, China had been reluctant to take risks by intervening in the Korean conflict.[53] Chinese troops had engaged UN troops in an evident effort to persuade the UN command to reconsider its goal of conquering the entire North. But it was only after China's vital interests were threatened – the security of access routes to China's industrial heartland – that China entered the conflict, when US troops had launched their home-by-Christmas offensive on the Yalu.[54] While the perceived threat to China's vital interest was China's cause rather than North Korea's, the Chinese were not afraid to shed blood on behalf of the North Koreans.

The relationship between North Korea and the People's Republic of China had not developed significantly before the Korean War,[55] but the entry of the Chinese army drastically transformed this situation. Indeed, Kim Il Sung learned a lesson from the war about the support he could expect from his two Great Power allies. Although Stalin approved the selection of Kim Il Sung as a national leader due to Kim's experience in the USSR and his training under Soviet commanders, the Soviet Union was not prepared to come to North Korea's aid and risk war with the United States.[56] But the fact that Sino-Korean friendship was sealed in blood in the Korean War has affected the policy makers in the US right up to the present. However, perhaps the most important result that the Korean War produced was that in its aftermath the North Koreans became capable of dealing with two major powers. As long as the Soviets and the Chinese were united, North Korea got help from both. When their relations deteriorated, however, the involvement of both in Korea imposed diplomatic problems but also provided manoeuvring room for the North Koreans to increase their independence.[57]

As far as the USSR was concerned, the US containment policy towards China increased Chinese dependency on the Soviets. Before the Korean War, Mao had set his mind on liberating Taiwan. By doing so, he had sacrificed all hope of normalizing ties with Washington for the foreseeable future. Liberating Taiwan was Mao's priority, but the Korean War made the attainment of that goal impossible. When the war ended, Mao returned to his priority task, clearly expecting Soviet support. However, Stalin's reluctance to provide direct and immediate assistance to the Chinese in Korea raised Mao's suspicions as to the reliability of the Soviet deterrent for China under the terms of the treaty. Stephan Kaplan concludes that the Soviets were anxious to avoid provoking a US attack on the Soviet Far East and worried that Soviet military intervention in North Korea would lead to a Third World War.[58]

It can therefore be argued that the Korean War finally brought each of the Great Powers to a shared consensus that the Korean peninsula presented a threat rather than an opportunity. The way the Korean War was terminated crucially shows the Great Powers' consensus; that the peninsula was primarily defensive, or at best a way of protecting the status quo to prevent any worsening of the situation in terms of their own security concerns. When the United States entered the Korean War in 1950, however, its motive was one of defending the Republic of Korea, but once the northern forces were routed, it changed quickly to the offensive. President Truman resisted MacArthur's recommendation to extend the war into China and use atomic weapons.[59] On 17 May 1951, the Truman administration adopted the recommendation of an important document (NSC 48/5) specifying that the ultimate objective was to seek by political as opposed to military means, a solution to the Korean problem.[60] The document explains that a political settlement should not jeopardize the United States' position with respect to the USSR, to Formosa, or to accepting Communist China's membership in the UN. China also showed unwillingness to risk a larger war, as did the Soviet Union. The Korean War thus made it clear that, for the Great Powers at least, the division of Korea satisfied their minimum security requirements.[61] For them, the status quo was preferable to incurring the risk of more widespread war, and this is why US and Chinese officers met with North Korean officials in Panmunjom to monitor the truce.[62] In exchange for acceptance of the ceasefire arrangement from President Rhee, who judged it to be against Korean unification, Secretary of State Dulles and South Korean Foreign Minister Pyun Yong Tae initiated the Mutual Defence Treaty on August 1953 in Seoul and signed it in October in Washington.[63] In addition to preventing another Korean War, the continuing US military presence in South Korea was obligated to serve what stability there was in the balance of the Asia–Pacific region and in particular, to protect Japan's security interests.

    The Korean War prompted the United States to terminate its occupation of Japan in 1951 and to transform its erstwhile enemy into an ally against the communist axis of the Soviet Union and China by signing a security treaty with Japan in 1952. China notably concluded a treaty of friendship and alliance with Soviet Union in February 1950 against the emergence of this US–Japanese alliance.[64] Following its defeat in World War II, Japan expected the United States to guarantee the security of both South Korea and Japan, but the US expected Japan to make a greater contribution to regional security in East Asia by significantly increasing its own defence expenditures. Unlike the pre-World War II situation, however, in the postwar era Japan treated Korea as an issue important only to its relations with the United States. Indeed, as long as the United States regarded its alliance with Japan as the cornerstone of its strategy in this region and as long as Japan regarded its relations with the United States as the axis of Japan's diplomacy, both sides of this axis could envisage Korea as part of their bilateral relations.[65] That explains why normalization between Japan and South Korea did not materialize until Japan was able to view its alliance with US as essential for its security. The process of reaching this point, however, was no easy road.

The San Francisco Peace Treaty and the US–Japan Alliance

Stalin's continuing concerns about Beijing's possible relations with the West was mirrored by the concerns of American policymakers, who were also worried that Japan would work for its enemies unless Japan's complicity could be assured after the signing of the San Francisco peace treaty. The Great Powers' sixth sense certainly appears to ring true if we examine in particular Japanese politics and diplomacy during the postwar years that preceded the 1960 US–Japan security treaty. The alliances between China and Russia and the US and Japan were pragmatic marriages in which each partner wanted desperately to trust the other since there were no pre-existing links strengthened by deep historical content and no long history of close cooperation or cultural affinity. While most American planners hoped to anchor Japan as firmly as possible to the anti-Communist West, the majority of Japanese aspired to at least nominal neutrality in the Cold War so that they could concentrate on their own economic recovery. This required a three-fold commitment from the Japanese: that they avoid large-scale rearmament, reach a peace settlement that included China and the Soviet Union, and prevent or minimize the presence of American bases on their territory.[66] Although a fervent anti-Communist, Shigeru Yoshida – the first influential Japanese leader in the postwar years – largely shared these sentiments. So how did subsequent Japanese leaders play a role in negotiating the security treaty, setting the course for Japan's subsequent foreign policy and ultimately determining its relations with the United States as the axis of Japan's diplomacy? The following pages examine the transition process of Japan's foreign policy from neutrality into becoming an important American ally in the region by looking at three key Japanese leaders – Shigeru Yoshida, Ichiro Hatoyama and Nobusuke Kishi.

Between 1945 and 1950, Japan experienced what Occupation Commander General Douglas MacArthur called a 'controlled revolution' – the partial uprooting of political, economic and social structures that had contributed to repression at home and aggression abroad. The revision of Japan's constitution was the most important single reform undertaken early in 1946 after the termination of the Pacific War. The document, which was drafted by the Americans and had to be translated into Japanese, stripped the emperor of temporal authority, enhanced the Diet's power, extended voting rights and declared the legal equality of women. Article IX, to Washington's later regret, forbade the creation of armed forces or the right of Japan to conduct war.[67] In the pre-war period, Yoshida had been a vocal supporter of good relations with the British Empire and the US. As Japanese ambassador to London during the 1930s, Yoshida viewed Japan's aggression in Asia with alarm. By 1945, he was joining those urging the emperor to negotiate an end to the war before a Soviet invasion or a leftist revolution.[68] It was this background that made him acceptable to the occupying forces. In the spring of 1946, Shigeru Yoshida formed his first postwar cabinet.

Despite Yoshida's positive attitude towards the Allies, he also believed that Japan's destiny lay in maintaining a close association with a politically and economically dependent China. Yoshida's knowledge of the Chinese classics was extensive and he was well versed in Chinese poetry.[69] Nevertheless, unlike many his pan-Asianist contemporaries, Yoshida believed that Japan's continental ambitions could not be realized in the face of British and American opposition. The Chinese, Yoshida firmly believed (on the basis of his long experience on the continent), were a pragmatic, materialistic and highly individualistic people. Throughout the postwar period he never abandoned his private conviction that after the Sino–Soviet alliance had collapsed, and when a Japan strong enough to negotiate with Beijing on an equal footing had freed itself from American tutelage, the two great East Asian powers could resume their natural historic relationship.[70]

Many of Japan's postwar achievements can be credited to Yoshida. The restoration of national sovereignty was a primary objective for most Japanese, and it was during Yoshida's first Cabinet (16 May 1946–23 April 1947), after the enunciation of the Truman Doctrine and MacArthur's call for an early peace settlement, that Japanese officials began to pay serious attention to the nation's future foreign and defence policies. As the relations between the superpowers – primarily the US and the USSR – gradually deteriorated, Yoshida decided that Japan's interests might be best served by forging a stronger but temporary alliance with the United States, involving the stationing of American forces in Okinawa and the Bonins.[71] In May 1950, he dispatched a secret mission to Washington to formally propose the conclusion of a bilateral security treaty. Yoshida's offer envisaged American bases in Japan acting to preserve the security of Japan in the Asian area.[72] President Truman appointed John Foster Dulles as special ambassador to begin negotiations on a peace settlement with Japan and set up some form of comprehensive defence arrangement. The main points of controversy in the negotiations were Japanese rearmament, participation in regional security arrangements and the question of an explicit American guarantee to defend Japan.[73] On 9 February 1951, Dulles and Yoshida reached agreement on several matters, including a provisional peace treaty, a draft collective self-defence agreement, a draft agreement permitting US/UN forces to utilize Japanese facilities in support of Korean operations, a draft status of forces agreement and an agreement regarding services and facilities that Japan would provide to American forces.[74]

Two important points were decided between the parties. From the American side the land, air and sea forces the US maintained in and around Japan for use anywhere in the Far East, and which were not required specifically to defend Japan, could be withdrawn at any time or used to suppress internal disturbances, and that the security treaty could be terminated only by mutual consent.[75] From the Japanese perspective, by using his own powers of negotiation, Yoshida was able to insist that economic recovery be prioritized and that Japan's troops would never become American surrogates charged with policing Asia.[76] It was not until the Korean War that the United States reversed its position of non-intervention, which had accounted for its absence in the Chinese Civil War. At that point, the Japanese archipelago became America's most crucial forward base for operations in the Far East, and a rearmed Japan became a pivotal link to US plans for Asian–Pacific regional security modelled on NATO guidelines.[77] For the Japanese, however, the Korean peninsula still continued to be seen as a 'dagger thrust at the heart of Japan': both a problem and an opportunity. But as far as Japan was concerned, the opportunity began to outweigh the problem. The Korean War was described by Yoshida as the 'Grace of Heaven.' Ambassador Robert Murphy also described the war as a 'godsend' for the Japanese, giving them the excuse and the ability to rebuild at maximum speed.[78] The procurement boom transformed Japan into one huge supply depot, which at the same time ruled out the initial worries of US planners that Japan could at any time place herself in a position of dependence upon the Communist-dominated mainland of Asia.[79]

More importantly, Japan finally got a chance to end the occupation and recover its national sovereignty, while Communist China's intervention in the Korean War precluded Japan's earlier strategy, resulting in Yoshida's acceptance of rearmament, the provision of foreign bases and constraints on Japanese ties to China. The San Francisco Peace Treaty was signed on the morning of 8 September 1951 between Japan and forty-eight other nations. The Soviet Union and two of its allies, Poland and Czechoslovakia, refused to accept the treaty. India and Burma refused to attend due to the exclusion of China. In order to convince the British to drop their demand that China participate in the peace settlement, neither the Communist nor the Nationalist Chinese regime was invited to the peace conference. Tokyo's residual sovereignty over the Ryukyu Islands was confirmed, but it had to surrender administrative control of these, and several other small islands, to the United States.[80] At 5pm the same evening, two US senators (Acheson and Dulles) and Yoshida signed a US–Japan security pact along with a subsidiary agreement that authorized US forces to use bases in Japan for Korean operations.[81] A few days before, the United States reached defence pacts with the Philippines, Australia and New Zealand, which formed the core of the American military presence in the Asia–Pacific region for the next quarter-century.[82]

However, Yoshida's plans for joint US–Japanese security differed greatly from the role envisioned by the US for Japan as far as the maintenance of international peace and security in the Far East was concerned, especially in the event of the hostilities on the Korean peninsula. Shortly before the occupation ended, Yoshida called for the transformation of the police reserve into something that resembled more of a a self-defence force. Japanese ground forces were located in thirty-seven bases near urban centres of domestic Communist strength and were organized into four army divisions.[83] According to Mun Hak Bong, President Rhee's private secretary and political advisor, the United States drew up plans early in 1950 to have a reconstructed Japanese army dispatched to the Korean peninsula in the event of possible hostilities, to encourage Japanese nationals to join the South Korean forces and to have the South Korean officer corps trained in Japan and a South Korean armaments industry developed on Japanese territory under Japanese management.[84] These projects flew in the face of historical colonial legacies. Yoshida, like many Japanese officials of his generation, disliked Koreans in general and despised South Korean president Syngman Rhee in particular. Rhee in turn frustrated Washington's attempts to promote a wider measure of politico-strategic cooperation between its two Northeast Asian allies due to his own intense anti-Japanese sentiments throughout the 1950s. Yoshida's fear of militarism, his deep prejudice against Koreans and his desire to focus on economic recovery ruled out not only the plans of the Joint Chiefs of Staff for Japanese military self-sufficiency but also any military cooperation with South Korea.[85]

Moreover, the Japanese peace treaty signed in San Francisco still left the Kuril Islands question unsettled, leaving friction between the USSR and Japan. On 11 February 1945, Roosevelt, Churchill and Stalin signed the Yalta Far Eastern agreements, pledging that in return for Russian entry into the Pacific War, the Soviets would regain Sakhalin Island and the Tsarist concessions in Manchuria, and that the Kuril Islands would be handed over to the Soviet Union.[86] However, awarding the Kurils to Moscow violated previous allied statements of principle. The Atlantic Charter and the Cairo Declaration authorized the victors to return only those territories Imperial Japan had seized by violence and greed. While this stipulation applied to southern Sakhalin, a Japanese war trophy, it clearly did not extend to the Kurils, which Tsarist Russia had ceded to Japan peacefully in 1875.[87] Given the development of US diplomacy and military strategy in Northeast Asia in the decade after Japan's surrender, American policy towards the Kurils must be seen in the context of the changing relationships between the United States, the Soviet Union and Japan. Although the completion of the peace treaty and the US–Japan security pact helped solidify the American position in the Pacific, improvements to the Soviet Union's military capabilities in Asia and developments elsewhere in the region dominated the concerns of the Joint Chiefs of Staff (JCS), especially regarding the overall balance of power in Northeast Asia.[88] The

Kurils might have served as a deterrent in the north, and postwar planners regarded the placing of an airfield on one of the islands as important not only as a hedge against possible Soviet aggression, but also because of its potential as an emergency station on the air route to Japan and other destinations in Asia.[89] Coincidently, Dulles had become disillusioned by the Yoshida government's policies, which showed a consistent reluctance to bear some responsibility for and a fair share of the common burden of the defence of the free world.

Following the July 1953 armistice in Korea military procurement orders declined, and in late 1954 Ichiro Hatoyama's Democrats replaced Yoshida's Liberals as Japan's governing party.[90] Although Hatoyama's past support for rearmament had pleased Washington, his interest in forging closer ties with the Soviet Union and expanding trade with the PRC spelled trouble. In the prevailing climate, the Kurils controversy could earn Moscow's good will with Japan. Unlike Yoshida, conservatives such as Ichiro Hatoyama and Mamoru Shigemitsu acknowledged the constitutional problems but attempted to solve them by instituting an amendment authorizing the creation of armed forces. According to these so-called revisionists, the complaint was that reliance on the Americans and the security treaty placed Japan in a subordinate position, preventing it from dealing on equal terms with China, Southeast Asia and the Soviet Union. By amending the constitution, Japan would be able to pursue a more independent foreign policy.[91]

The 1951 security treaty gave the United States the right to position its military forces in and around Japan without imposing any legal responsibility upon the US to defend Japan. Furthermore the United States, while assuming no specific defence responsibility, was theoretically allowed to use its forces in Japan for purposes not directly related to the defence of Japan. If the purpose was to contribute to the maintenance of international peace and security in the Far East, the United States was thus freely able to use Japanese bases for defensive operations.[92] After the Korean armistice had been signed in 1953 and the French position in Indo-China collapsed in 1954, the Eisenhower administration began to fear that its entire Asian–Pacific strategy had been jeopardized. Dulles and Admiral Radford (the chairman of the JCS) began discussing a plan to intervene in Vietnam in association with forces from America's Asian and Pacific allies. Around this time, Japanese–American differences over the controversial question of overseas service for the self-defence forces surfaced in August 1955.

Coincidently, as of 1955, Japan remained technically in a state of war with the Soviet Union. After refusing to sign the San Francisco treaty, Moscow vetoed Japan's membership in the United Nations, continued to detain several thousand Japanese soldiers captured in the last days of World War II, and retained four small islands just north of Hokkaido which the Russians had seized in August 1945.[93] In early 1955, as part of a new approach to the Soviet Union, Foreign Minister Mamoru Shigemitsu undertook to conclude a peace treaty with Moscow that would return Etorofu, Kunashiri, Shikotan and Habomais to Japan. By focusing its attention on the four islands closest to Japan, the Japan's Foreign Ministry asserted its claim to the so-called Northern Territories on the grounds that Japan's sovereignty over Etorofu and Kunashiri had been recognized by Russia in earlier negotiations. Moreover, the islands had always been administered as a part of Hokkaido, which meant that they were not legally considered part of the Kuril Islands. As for Habomais and Shikotan, the Japanese argued that these islands were not even a geographical part of the Kurils.[94] Unexpectedly, Moscow softened its position, allowing the return of Habomais and Shikotan in exchange for Japan's cession of Etorofu and Kunashiri. The Russians' intention, however, was driven by the hope of detaching Japan from its security treaty with the US. In response to this, Dulles expressed the fear that '… if the Soviets returned any of the Kuril lands, the United States would at once experience heavy Japanese pressure for the return of the Ryukyu Islands.' To avoid this, he persuaded Eisenhower to discourage the Japanese from compromising with the Soviets.[95] When Shigemitsu went to Washington to discuss the Far Eastern situation and press the case for revision of the 1951 security treaty, Dulles countered Shigemitsu's complaints about the inequalities of the 1951 arrangement with the argument that from Washington's standpoint, the treaty was an unequal one. Japan was under no obligation to come to America's aid in time of crisis.[96] Newspapers reported that the Secretary warned Shigemitsu that if the Japanese exceeded the terms of the 1951 peace treaty and ceded territory to Russia, the US would seek similar advantage from Japan.[97] This meant that Dulles would request the cession of Okinawa to the United States.

Faced with such pressure from the US, Hatoyama was no longer able to accept Moscow's offer, and the two sides settled for a minimal agreement. In a joint declaration issued on 19 October 1956, the Soviet Union and Japan announced an interim settlement terminating their state of war, exchanging ambassadors, returning Japanese prisoners of war, restoring fishing rights and endorsing Japan's admission to the United Nations.[98] The Hatoyama administration had given the US its first challenge of the postwar era, insofar as Japan's efforts to improve ties with its Communist neighbours had aroused American fears and provoked a strong US reaction. In November 1955, the main conservative groups merged to form the Liberal Democratic Party (LDP), financed by big business, committed to the alliance with the United States, and strongly supported by American diplomats. After Dulles blocked Hatoyama's attempt to forge a more independent foreign policy and preserved the security treaty, no further developments took place until Prime Minister Nobuske Kishi's visit to Washington in June 1957.

Kishi, a former bureaucrat who had worked to develop Japanese industry in occupied Manchuria as a minister in General Tojo's wartime cabinet, was a statesman of considerable ability. He established himself as the foremost Japanese exponent of active co-operation with United States Far Eastern policies. He consistently opposed the development of ties with the People's Republic of China, and considered that the Communist Revolution had destroyed Japan's position on the continent forever.[99] Kishi may have been a nationalist, but he was also acutely aware of American power. His vision was of a rising new Japan in a firm partnership with the United States – a partnership that would enable Japan to be a major player in Asia.[100] In order to achieve his ambitions, however, Kishi also hoped to put US–Japanese relations on a more equal basis by revising the security treaty. In particular, unlike former prime ministers, he was the first Japanese statesman to consider that the US–Japan security alliance not only provided security for Japan but also made Japan the US's principal Asian political ally. As such, he willingly contributed to the Republic of Korea's military defence against the communist bloc.

The revised treaty ruled out American military intervention to suppress Japanese domestic disorders and gave Japan the unequivocal promise of American protection against attack that Dulles had refused to extend to Yoshida in 1951 (Article V). The upshot of this was that Japan involved itself for the first time in a formal commitment to contribute to the defence of American installations, although this required Japan to maintain and develop its capacities to resist attacks within its own territory (Article II). Undoubtedly a significant change was that the treaty allowed either side to terminate the agreement after ten years. Perhaps the most problematic features of the new treaty were the provisions for prior consultation before combat deployment of US forces from Japanese bases or the introduction of nuclear weapons (Article IV), which allowed Japanese governments to veto any American requests to bring nuclear arms into Japan.

This may have looked good on paper, but empirical evidence suggests that Japan's ability to exert any influence over the deployment of US forces in its territory was always negligible. Japan was unable to control the character of US equipment or American operations beyond Japan, let alone the general development of American Far Eastern policies.[101] For example, in May 1981, in an interview with the Mainichi Shinbun, former United States ambassador to Japan Edwin O. Reischauer revealed that a secret verbal agreement, concluded at the time of the 1960 security treaty negotiations, permitted Washington to bring nuclear weapons freely in and out of its bases on the archipelago.[102] In addition, the interpretation of the treaty left numerous loopholes that enabled US forces to take action outside Japan without prior consultations with Japan. Although the treaty was limited geographically to territories under Japanese administration, Japan was committed in some degree to cooperating with American policy in the Far East as a whole (Article VI). Importantly, from the Japanese perspective, Vietnam was not included within the scope of the Far East, but after heated debate both inside and outside the Diet, the government retreated to the position that the Far East was intrinsically a vague area and as such it was not appropriate to discuss which areas were included and which were not.[103]

Despite his strong interest in a more militant anti-Communist alliance, Kishi was unable to negotiate a genuine bilateral military alliance with Washington, not so much because of opposition from the Socialist and Communist parties, but because of resistance from within the Liberal Democratic Party itself. The massive popular anti-treaty movement of 1959–60 strengthened the position of Japanese Conservatives who stressed the fact that Japan and the US were different societies with different cultural backgrounds. Indeed the majority of Japanese believed that the US–Japan security treaty was a pragmatic marriage, in which each partner strove desperately to trust the other in the absence of any deep historical context or cultural affinity. Few expected the treaty to continue until the present day – in fact, Kishi made no secret of the fact that he hoped to see the development of military ties between Japan and the countries of non-Communist Asia.[104] Coincidentally, this cause became particularly close to his heart after the military coup d'état that brought Major-General Park Chung-Hee to power in Seoul in May 1961, just twelve months after Kishi's own resignation as Prime Minister of Japan. Park Chung-Hee, a graduate of the Japanese Military Academy in Manchuria and a former officer of the Imperial Japanese Army, began a programme of political and economic reconstruction under strong military-bureaucratic control, a situation that was reminiscent of Kishi's own experiments in Manchuria.[105]

Exclusionary Policy toward the Zainichi in the Post-World War II Era (1945–64)

With the end of the Pacific War and the destruction of Japan's militaristic ideals, Japanese leaders upheld the principle of non-interference in Korean domestic affairs for many years. Most Japanese leaders who served from 1945 until 1964 unanimously acknowledged that colonization policies had completely failed to reconstruct a Japanese national consciousness amongst Koreans. In addition, given its weakened position and apparent lack of self-governing abilities, the entire Korean peninsula was forced to exist under the trusteeship of two Great Powers – the US and the USSR – immediately after the war. Given this situation, the foreign policy of the Japanese elites was reflected in the Great Powers' geopolitical strategies in the region. According to my argument, which is consistent with Mylonas' nation building theory, the main independent variable – that no interstate relations existed between the host state and the external homeland – coupled with foreign policy preferences made it easy for the Japanese elites to select an exclusionary policy toward the Zainichi diaspora (**Configuration 3**). As I explained in the previous section, the Japanese elites' foreign policy had gradually shifted from preserving the status quo towards a more revisionist outlook, a process that ultimately resulted in the formation of the US–Japan alliance. Having outlined the broader issues at play in East Asia above, we can now move on to consider their effect on the logic underlying the policies followed by the Japanese ruling political elites toward the Korean diaspora in Japan – the Zainichi – a logic that reflects the interstate relationships between Japan and the divided Korea as well as Great Powers' geopolitical strategy in the East Asia.

By the end of World War II, there were an estimated 2.4 million Koreans living in Japan. The official repatriation programme, overseen by the Supreme Commander of the Allied Powers (SCAP), began in December 1945, by which time tens of thousands of Koreans had already converged on Japan's southern ports to await transportation. By the end of 1945, more than 1.3 million Koreans had returned home.[106] In keeping with the Cairo Declaration of 1943, which had unequivocally stated that Korea would in due course become free and independent, the Occupying Forces' Basic Directive of 28 November 1945 declared Koreans and Formosans to be liberated people.[107] Repatriation from Japan to Korea and China was voluntary, although it had been generally assumed that almost all former colonial subjects would take the opportunity to go back to their place of origin. However, by the early months of 1946 some of the repatriation ships from Senzaki and other ports were leaving half empty.[108] Those Koreans who desired to remain in Japan, around 650,000 in number, were mainly long-term residents with pre-war roots in the country. These people were mainly uprooted migrants who had nowhere to go, although a variety of factors prompted their decision to remain. In November 1945, GHQ (General Headquarters) had imposed a ¥1,000 limit on the amount of property returnees could take with them. In addition, floods, epidemic diseases, rice riots, lack of housing or jobs, prejudice against repatriates, political instability and a homeland occupied by US and Soviet troops convinced many Koreans to stay in Japan.[109] By the end of 1946, the rate of Korean repatriation had slowed to a trickle, while the numbers of people who were immigrating illegally to Japan from Korea was increasing rapidly. Those who had returned to Korea faced a changed homeland in which only Japanese or Japanese collaborators remained in key positions and Korean citizens had less freedom south of the 38th parallel than they had enjoyed in postwar Japan.[110]

As soon as the pre-war colonial subjects that they were told that they were a liberated people, the SCAP began to exercise criminal and civil jurisdiction over United Nations nationals in Japan. However, questions were raised in terms of the obscure status of jurisdiction over Formosans, Koreans and neutral foreigners. Then, regarding Formosans and Koreans, the occupation forces in Japan were told:

> You will treat Formosan-Chinese and Koreans as liberated people in so far as military security permits. They are not included in the term 'Japanese' ... but they have been Japanese subjects and may be treated by you, in case of necessity, as enemy nationals.[111]

The Japanese government, having disenfranchized Korean and Formosan residents in late 1945 and written them out of its constitutional draft in early 1946, now sought to expel as many as possible. Meanwhile, the SCAP intended that everyone who had been a Japanese national on the day before Japan's surrender would remain a Japanese national, unless they manifested an intention to adopt another nationality. SCAP's intent was that Koreans in Japan should be considered as retaining their Japanese nationality until such time as an established Korean government accorded them recognition as Korean nationals.[112]

The policy toward these Koreans remaining in Japan precisely reflected the Japanese elite's strong desire for economic recovery, especially in the light of Prime Minister Yoshida's perception of Koreans as well as his status quo foreign policy. Yoshida certainly considered Koreans as being ungrateful for the benefits bestowed upon them during the thirty-six years of Japanese colonial rule – a view that was most emphatically not shared by the Koreans. For him, the history of the pre-war era was regarded as disastrous involvement in conflicts with China over spheres of influence in Korea; the peninsula was no longer seen as a dagger pointed at the heart of Japan but as a one way road to catastrophic entanglements on the continent.[113] Although President Truman attempted to involve Japan in regional military organizations after the Korean War, Yoshida was particularly hostile to any suggestion of a military role on the peninsula. Korean President Rhee responded by openly manifesting intense anti-Japanese sentiments until he resigned in 1960.

Under such conditions, policy making toward the Koreans in Japan was certainly exclusionary. Instead of treating them as either Allies or enemies, Japanese pejoratively referred Koreans as third-country nationals (*daisan-kokujin*).[114] Two thirds of Koreans were unemployed and turned to black-market activities and the illegal distillation of liquor to support themselves. SCAP's refusal to treat Koreans as UN nationals heightened tensions between Japanese and American authorities and the Korean diaspora.[115] In November 1946, the Japanese government announced that Koreans would have to pay a special capital levy or war tax. On 20 December, some 30,000 Koreans rallied in front of the Imperial Palace to protest at the loss of voting rights, the war tax and the imposition of Japanese-national status.[116] The protestors were tried by an Eighth Army military tribunal and deported en masse to southern Korea on the orders of Eighth Army Commander Robert Eichelberger. The use of super ordinate occupation courts to deport the protestors signalled the starting point of a hardening of American attitudes toward the Korean diaspora.[117] As early as the spring of 1946, MacArthur's headquarters had concluded that the presence of a restless, uprooted Korean minority in Japan, disdainful of law and authority, was a serious obstacle to the success of the occupation.[118] In June 1946, the Home Ministry, via the Central Liaison Office, asked SCAP for sweeping powers to deport Korean 'troublemakers'. Although GHQ rejected the proposal on the grounds that it was discriminatory and was not designed to be universally applied to all foreign nationals, American and Japanese authorities began at that point to consult closely on what they termed the Korean problem.[119]

Finally in 1947, the Alien Registration Ordinance (ARO) was drafted for administrative and control purposes by Korean specialists in the Government Section (GS) and their counterparts at the Home Ministry.[120] In order to avoid parliamentary debate on the issue, the ARO came into force the day before the new Constitution went into effect. Ironically, it seemed to be one of the final edicts issued under the old Imperial decree, which had been a powerful instrument of the Governor-General of Korea, *Chosen Sotokufu*, during the colonial era.[121] The 1947 ARO required Koreans and Formosans to register from the age of 14 and to carry an alien passbook at all times.[122] Failure to comply was a deportable offence, a stipulation that implicitly undermined the legal status of former colonials as Japanese nationals.[123]

Japan's exclusionary policy toward Koreans, which was shared by American occupation policies, also denied the Koreans autonomous schools, since a separate education system would have fostered a strong and distinctive national identity, intensified ethnic antagonism and ultimately complicated the task of occupation.[124] Following instructions from SCAP in January 1948, the Ministry of Education ordered Korean children to attend the same schools as Japanese from 1 April, the start of the new academic year.[125] This policy was based on the American assumption that Koreans would assimilate themselves into Japanese society and that the primary agent of absorption should be the schools, as was happening in the United States. The SCAP, however, was met by a challenge to its authority from the League of Korean Residents in Japan (*Choren*), which had been created in mid-October 1945 to defend liberated nationals and other groups, and which adamantly opposed the new policy.[126] In March and April 1948, Koreans across Japan rose up in protest after the Japanese government began to implement the order handed down to them by SCAP to close Korean ethnic schools. One of the famous protests took place in Kobe on April 24 when Koreans stormed the Hyogo Prefecture offices in an attempt to persuade the governor to rescind the order to close the four local Korean schools.[127] A similar protest took place in Osaka on 23 April. SCAP and the Japanese government reacted harshly to the Korean actions in these prefectures, with police arresting thousands of Koreans and inflicting strict punishments on the incidents' ringleaders.

Faced with strong resistance in the form of the Kobe-Osaka demonstrations, MacArthur finally asked the Diplomatic Section to prepare a staff study on the status of Koreans in Japan. Richard B. Finn was appointed by ambassador Sebald to draft a new plan on the nationality of Korean residents. According to statements from Jules Bassin and Douglas MacArthur in August 1948, 'the large Korean group in Japan, which is for the most part inassimilable … and the source of dangerous friction with the Japanese constitutes a strong element of instability in the Far East and the cause of unfavourable propaganda directed against the United States.' MacArthur concluded that 'if the treatment now accorded United Nations nationals or other foreigners in Japan were extended to Koreans, the position of Koreans in Japan would become further entrenched in direct conflict with SCAP policy to encourage their return to Korea.'[128] In May 1949, however, SCAP reports insisted that

Koreans should be offered three choices: repatriation, the acquisition of full Japanese citizenship, or the retention of permanent alien status. 'In this way', the report stated, 'Koreans in Japan will eventually be divided into distinct legal groups and so divided will be subject to better control, both for purposes of repatriation and absorption.' However, Neither SCAP, Japan, nor the Republic of Korea intended to give Koreans a choice of citizenship. Finn was therefore forced to conclude that 'the problem of legal status should be resolved by Seoul and Tokyo after the occupation.'[129]

Yoshida's policy of choice was repatriation or forceful deportation. In July 1949, Yoshida's aide Jiro Shirasu visited the US Diplomatic Section and proposed a drastic solution to Japan's Korean problem, calling for the deportation of five or six hundred thousand North and South Koreans.[130] In late summer, Yoshida formally petitioned MacArthur to authorize this project, complaining that the Koreans were mostly communists or communist sympathizers, criminal elements, or parasites who contributed nothing to the economy. He proposed deporting all 650,000 remaining Koreans to South Korea at government expense, but MacArthur rejected the unilateral use of force to remove them,[131] although his stance definitely refused to provide full Japanese citizenship to Koreans or even the retention of permanent alien status. The final suggestion that Finn recommended as a solution for the post-colonial settlement for Japan's Korean diaspora group was negated by Yoshida's strong objection to Korea being a signatory to the San Francisco Peace Treaty in 1951. In April 1951, following MacArthur's recall, Yoshida and Dulles met to discuss the question of whom to include in the peace process. President Rhee of the Republic of Korea insisted that his country should be represented at the treaty conference. Dulles initially agreed, since he viewed Seoul's participation as a first step towards normalizing relations between Japan and South Korea whilst helping the US to build up the prestige of the Korean government. To Dulles' surprise, Yoshida replied that if South Korea became a signatory, Koreans in Japan (most of whom were allegedly Communists) would acquire all the compensation and property rights accruing to allied nationals. Yoshida pointed out that the government still wished to send almost all Koreans in Japan back home since the government had long been concerned about their illegal activities.[132] In addition, London had surprisingly also raised objections to the inclusion of the Republic of Korea as a signatory, on the grounds that it was not a member of the Far East Commission (FEC).[133] The primary motive for London's objection was British ire over American opposition to Chinese participation. The Nationalist government, not the People's Republic, represented China on the Far Eastern Commission. If Beijing, a non-FEC member, could not attend the San Francisco Conference, then, London reasoned, neither should Seoul. The combination of Japanese and British opposition thus eliminated Seoul as a co-signatory to the American draft.[134]

With the Republic of Korea no longer a party to the peace, the San Francisco settlement was able to ignore the contentious issue of Koreans in Japan. On 28 April 1952, the day the San Francisco Peace Treaty came into force, Japan's Justice Ministry unilaterally stripped Koreans and Chinese of their Japanese nationality. Simultaneously, the government enacted Law 126, which permitted Koreans living continuously in Japan from 2 September 1945 through to 28 April 1952 to remain in the country for the time being, pending the ultimate resolution of their legal status. As aliens, however, Koreans and Chinese became immediately subject to the Immigration Control Law (November 1951) and the more powerful Alien Registration Law, which closely mirrored both wartime Japanese migration control and the US McCarran-Walter Act.[135] The New Immigration Control Ordinance defined deportable foreigners as including people convicted of crimes other than minor offences. The postwar Japanese law could also extend to the deportation of foreigners on political grounds. Reflecting the Cold War fears of invasion by infiltration, deportation rules in Japan particularly emphasized sabotage or the destruction of public property.[136] Indeed, Koreans and Communists were targeted by the Subversive Activities Prevention Law of 21 July 1952, while the alien registration system made life even more difficult for stowaways. Life was to become harder again in 1955, when fingerprinting was introduced as part of the registration process.[137] Without registration documents, aliens had no way of obtaining rations, jobs or medical care and lived in constant fear of arrest and deportation.

Prime Minister Yoshida's enthusiasm for mass deportation could not be brought to bear in full during his administration, although it failed to provide an immediate solution to the legal status of Koreans residing in Japan. However, during the next administration under Prime Minister

Hatoyama, the issue of repatriation again surfaced. The two key independent variables – foreign policy preference and absence of interstate relations between host country and homeland – conform to the argument presented in this book. According to Mylonas's theory, a host state is likely to exclude a non-core group when the state has revisionist aims and an enemy is supporting that non-core group.[138] My argument is in full accord with this. Mylonas defines exclusionary policies as policies that aim at the physical removal of a non-core group from the host state. Policies under this category include population exchange, deportation, internal displacement or even mass killing.[139] Although leaders such as Yoshida and Hatoyama had no interest in normalizing relations with South Korea, their foreign policies were different. Yoshida's foreign policy favoured the international status quo by focusing only on domestic economic recovery, whilst Hatoyama supported a revisionist philosophy by manifesting an interest in forging closer ties with the Soviet Union and expanding trade with the PRC. From 1955, Hatoyama's revisionist foreign policy began to open economic links to some of Japan's communist neighbours, which allowed the numerical reduction of Koreans in Japan in other ways – not by deportation, but by encouraging voluntary repatriation to North Korea. Two programmes of repatriation of supposed 'enemy' populations such as Koreans and Chinese took place during the Hatoyama administration, both of which came about because of his revisionist foreign policy and strict internal migration control.

According to Tessa Morris-Suzuki, the agreement between Japan and the Communist regime in mainland China to exchange returning Japanese for departing Chinese influenced the plan to send large numbers of Koreans from Japan to Communist North Korea.[140] Since Japan had political relations with Chiang Kai-shek of the Republic of China (ROC), this was the only place to which it could officially deport Chinese arrested for illegal entry or for other offences. But some of those awaiting deportation in Hamamatsu Detention Centre came from the mainland and risked persecution and separation from their families if deported to Taiwan. In order to resolve this situation, the Chinese government asked Japan to repatriate Chinese residents who wished to return to the PRC.[141] The origins of the repatriation plan were traced back to an agreement reached between the Red Cross Societies of China and Japan in 1953 to bring back to Japan some of the thousands of Japanese people who had remained in mainland China after Japan's defeat in 1945. Similarly, early in 1955, an informal arrangement was made with ninety of these detainees, under which they would be released from detention as long as they promised to board a Red Cross repatriation ship.[142]

In the mid-1950s, although the Japanese economy moved from postwar recovery towards high growth, there were many Koreans in Japan who remained impoverished and were either unemployed or employed in marginal areas of the economy such as day labouring, running pinball parlours (*pachinko*) and garbage collecting. According to the Yomiuri Newspaper, in the late 1950s, seven times as many Koreans as Japanese received Livelihood Protection (*Seikatsu Hogo*), a cost amounting to 1.8 billion yen annually.[143] Despite the fact that the majority of the Koreans in Japan came originally from the southern part of Korea, the South Korean government showed no enthusiasm for accepting a further swathe of returning Koreans from Japan due to its own struggles with poverty, high unemployment and its suspicion about the loyalty and ideological inclinations of the Korean diaspora in Japan. MacArthur reported in a telegram to Washington supporting this argument; Foreign Minister Fujiyama explained that:

> One of the principal factors favouring repatriation has been that of internal security. With increasing agitation from pro-Communist Koreans in Japan, the problem of controlling demonstrations and preventing riots has been a special concern for public safety authorities and the Justice Ministry ... The burden of destitute Koreans on Japanese government institutions at all levels totals 2.5 billion yen annually.[144]

In terms of foreign policy, the transition of power in Tokyo from Yoshida to Hatoyama in December 1954 caused serious anxiety for the United States. The Hatoyama administration's revisionist aims allowed Japan to pursue offers by the Soviet Union and China to improve trade and even diplomatic relations.

The mid 1950s also was a transitional stage for North Korea, during which Kim Il Sung attempted to consolidate his power by withdrawing from the Soviet orbit. During this time, he began to construct his own loyal party and build power by ousting Soviet-Koreans who retained Soviet citizenship and membership in the Communist Party of the Soviet Union, as well as Yenan Koreans who had returned from Yenan, China.[145] Coincidently, the pro-North Korean organization, the General Association of Korean Residents in Japan (*Chongryun*) was also founded in 1955. *Chongryun* officially avoided involvement in Japanese politics and encouraged its members to identify themselves as citizens of the DPRK. The Yomiuri newspaper reported Kim Il Sung's announcement on the *Chongryun's* goal:

> The fundamental goal was clearly targeted at organizing the Korean compatriots in Japan around the North Korean state and leadership and to work for Korea's reunification. It should not be aimed at overthrowing the Yoshida or Hatoyama cabinet ... The revolution of Japan should be taken by Japanese people, not by 600,000 Koreans in Japan ... if Koreans in Japan want to repatriate to our country, we will allow them to do so as soon as possible. These problems should be conducted based on mutual peaceful co-existence between Japanese government and DPRK.[146]

The opportunity to ride on a wave of revisionism coincidently favoured both Japan and North Korea in terms of their own aims. On 1 July 1955, Inoue Masutarô (a former Ministry of Foreign Affairs expert on communism) became the Director of the Foreign Affairs Department of the Japanese Red Cross Society. He was the main protagonist of the programme of repatriation from the beginning, arguing in a report of the Japanese Red Cross Society that:

> The ideal solution would be for the Japanese government to deport all the Koreans residing in Japan to Korea ... just as the Polish government deported all Germans to East Prussia, since this would eliminate the seeds of eventual disputes that might cause a trouble between Japan and Korea in future.[147]

Six months after the founding of *Chongryun*, the Fifth Section of the Asian Bureau of Japan's Ministry of Foreign Affairs produced a secret 'Plan for Resolving the Problem of Sending Volunteers for Repatriation to North Korea.' This plan aimed to establish a repatriation agreement between Japan and North Korea similar to the one signed between Japan and the People's Republic of China. As in the Chinese case, the agreement was signed not between governments, but between the Red Cross Societies of both countries.[148] The proposed procedures also spelled out the structure of the repatriation which the ministry envisaged for the newly established *Chongryun*. The exodus was to be carried out on the basis of a register of volunteers for repatriation that had been drawn up by *Chongryun*. The Japanese Red Cross Director of Foreign Affairs also pointed out that the ruling Japanese Liberal Democratic Party had recently taken a decision to start a movement to support the repatriation of the Koreans to North Korea.[149]

By this time, however, two possible obstacles to this scheme had become apparent. Firstly, it was clear that strenuous protests by South Koreans would become a major barrier to any repatriation agreement with the North, since Rhee's South Korean government clearly insisted that all Koreans were its citizens and repatriating any of them to a section of the country that he considered was temporarily under enemy occupation was simply out of the question.[150] Secondly, the United States was a major ally of both Japan and South Korea, and as the leader of the anti-communist free world was certain to look with some concern on any large-scale migration of people from a non-communist to a communist country.[151] In order to overcome these concerns and demonstrate the project was entirely apolitical and had humanitarian aims, the Japanese Red Cross argued that the International Committee of the Red Cross (ICRC) should be brought into the process and become a central participant in the repatriation.

Throughout 1957 and 1958, Japan continued to request ICRC commitment to the project. The South Korean government, however, quickly expressed outrage when it learnt of the repatriation project, threatening not only to break off normalization talks with Japan but also to use the Japanese fishermen detained in Busan as leverage to prevent the repatriation from going ahead.[152] Meanwhile, by the late 1950s, the US had other strategic concerns in East Asia whose importance far outweighed the significance of repatriation. Although the US government acknowledged that the repatriation might hinder the normalization of relations between Japan and South Korea, Washington contented itself with pushing for a clearer and more active ICRC role in the project and took no firm steps to intervene or prevent the accord from being signed.[153] From the US perspective, the security treaty with the Japanese government was a more important concern. After Prime Minister Hatoyama resigned following the agreement with Moscow in December 1956, Kishi Nobusuke was selected as Prime Minister in 1957. Since Kishi was a firm supporter of the US–Japan security treaty, the second half of 1959 was no time for the United States to oppose the single policy the Kishi administration had continuously pursued since the previous administration. The US concerns were focused on the fact that any replacement for Kishi was likely to be less sympathetic to American concerns.[154]

In September 1958, the North Korean leader Kim Il Sung publicly announced his country's decision to welcome all Koreans from Japan who wished to return to the bosom of the fatherland, saying that the North Korean government would offer repatriates from Japan free housing, welfare and education.[155] North Korea's wholehearted commitment to mass repatriation in 1958 was based not on humanitarian grounds, but on its own calculations of self-interest. Firstly, repatriation would help fulfil the labour demands for North Korea's own economic development plans. After a purge of the army in 1958, the remaining Soviet-Koreans and Yenan Koreans had been removed from the party. Nevertheless, North Korea had undertaken on a broad scale the Flying Horse Campaign, a project that was similar to China's Great Leap Forward, in an effort to expand production 'as rapidly as a horse that travels one thousand *li* per day'.[156] Secondly, the repatriation project would serve the North Koreans' desire to disrupt the movement towards normalization between Japan and South Korea. Finally, North Korea believed that the US would oppose a mass repatriation of Koreans to the DPRK and this would eventually damage Japan–US relations at a crucial moment in their evolution.[157]

The DPRK's policy shift towards the mass repatriation project was extremely successful. In November 1958, the Zainichi Korean Repatriation Cooperation Society was established by prominent figures including politicians from all sides of Japanese politics and headed by former LDP Prime Minister Hatoyama. The Japanese media also expressed strong support for the large-scale departure of Koreans to the DPRK. The Yomiuri newspaper explained the rapid development by exemplifying the speed of construction of housing, with one of their reporters stating:

> I have stayed in Pyongyang for almost 10 days, but the incredibly rapid construction rush for housing makes me confused to recognize where I walk. Most people emphasised that a new house would be built up almost every fifteen minutes ... Now I can understand that it is true.[158]

In this environment, the number of people volunteering for immediate repatriation to North Korea soared. Finally the ICRC had sufficient faith in the commitments given by the Japanese government to endorse the repatriation accord, which was then officially signed in the Indian city of Calcutta on 13 August 1959. The first repatriation ship left the Japanese port of Niigata for Cheongjin in North Korea four months later, and between that time and the termination of repatriation in 1984, 99,340 people left Japan to start new lives in North Korea.[159]

Although Japan's final goal was to repatriate 600,000 Koreans to North Korea, the number of repatriates was already in decline by November 1960. The Yomiuri newspaper reported on 12 December 1960:

> During the time periods following the first departure, at least more than 1,000 Zainichi Koreans had departed for North Korea annually, yet much smaller numbers, only 668 people, left Japan for North Korea this September. The Japanese Red Cross explains the reasons for this as follows: Firstly, North Korea's approaching harsh winter weather, secondly, most of the Koreans who were eager to repatriate had already left, and thirdly, since the end of year, it was difficult to sell or release the property they left behind in Japan … and so on.[160]

As a matter of fact, both the Japanese and the North Korean authorities quickly became aware that within months of the departure of the first repatriation ship, many of the newly arrived migrants from Japan were suffering distress and hardship in the DPRK. For example, in January 1961, a meeting between the Japanese Red Cross and government officials was told that 'according to letters received by those who have repatriated already, it seems that life in the North is really quite hard'; nine months later, any uncertainties had disappeared and an ICRC delegate in Japan could write of the undeniable truth in reports about the harsh existence of returnees in North Korea.[161] Among the victims there was a family of Kim Kyu Il, a first generation Korean who in 1982 founded the *Zainichi doho no seikatsu o kangaeru kai*, or non-profit ethnic Korean organization, which provided a forum for all Korean residents to create a unified vision for the Zainichi Korean community. Kim Kyu Il's family departed for North Korea, leaving him behind, but he subsequently found himself unable to contact them. In fact, the repatriated Koreans had been treated with great suspicions by the DPRK authorities since they started to move back to North Korea, and they ended up comprising a large percentage of the prison camp population. Kim Kyu Il was supposed to leave for North Korea at a later date, after finishing his studies as a promising student of Waseda University. But what he didn't know was that North Korea wanted the repatriated Koreans to provide a labour force to help rebuild a country that had been devastated by the Korean War. Educated Zainichi Koreans like Kim were unwelcome, and many were executed by the North Korean authorities. This was the fate that befell Kim's family, leaving him alone in Japan with no immediate news of his family's fate. When I spoke to him, I asked him who he thought was responsible for the tragedy. He answered that was his own responsibility, since those who recommended repatriating to North Korea had already left *Chongryun*. He had been one of the people in *Chongryun* who had believed that North Korea was an ideal place for the Zainichi – indeed for all of Koreans – and advised his parents to repatriate to the North. He felt unable to abstain from North Korea's utopian dream, the idea of 'unification' that was the slogan for the pro-North Korean diaspora association. Since those days, he has left the membership of *Chongryun*, got married and named his own son Tong Il which literally means 'unification' in Korean.[162]

By the mid-1960s, most Zainichi Koreans had become increasingly aware of the disappointing reality of life in North Korea. Some North Koreans also began to express unease and suspicion towards the alien values that the returnees were bringing with them into North Korean society.[163] Since Zainichi Koreans had grown up in a capitalist society for a long period of time, they became targets for discrimination, finding themselves as much an ethnic minority in their putative homeland as they had been in Japan. Furthermore, a considerable number of returnees had fallen victim to political purges. As noted above in Kim Kyu Il's experience, the families left behind in Japan also suffered separation from their relatives. Half a century after the start of the repatriation, many are still struggling to help their repatriated family members survive by sending them money and goods.[164] Many of the letters from repatriated Zainichi Koreans expressed the desire for eventual Korean reunification, closing with the words 'if Korea were unified someday, we would be meeting again.'

When Prime Minister Kishi and Foreign Minister Fujiyama met the US Secretary of State in Washington on 19 January 1960, three days after the revised Security Treaty was signed, Kishi took the opportunity to express his appreciation for the US attitude towards the repatriation of Koreans to North Korea.[165] Five months later, on June 23 1960, when the renewed United States–Japan Mutual Security Treaty was ratified by the Japanese Diet, Kishi announced his resignation, cooling the political temperature in Tokyo. The next Prime Minister, Hayato Ikeda, shifted attention away from contentious foreign policy issues to domestic concerns. Indeed, now more than any other ally, Japan's stability rested on the United States not only as its most important source of industrial raw materials and largest single market but also for leadership in fostering liberal trade policies throughout the free world and particularly among the industrialized nations of Western Europe.[166] In December 1960 MacArthur offered this assessment of Japan–US relations:

> The American market is essential and as long as Japan's daily bread depends largely on our cooperation and friendship, we do not believe a majority of conservative Japanese people will wish to chase the Communist rainbow.[167]

Conclusion

This chapter has highlighted the absence of any normalization treaty between Japan and the two Koreas, an absence that has allowed Japanese elites to pursue exclusionary policies towards the Zainichi diaspora. From the perspective of US strategy, Seoul–Tokyo reconciliation was prerequisite for what John Foster Dulles called the Pacific 'anti-communist arc'.[168] A normalization treaty between Japan and South Korea was an indispensable alignment, one which served the longer-term American enthusiasm for consolidating its Northeast Asian axis among asymmetric alliances. During the American occupation, the South Korean government had been allowed by SCAP to install what was called the 'Korean Mission in Japan,' accredited to SCAP in December 1948.[169] Under US initiatives, bilateral negotiations between Japan and South Korea had begun in Tokyo in October 1951, but the Korean Mission remained in Japan even after the conclusion of the US–Japanese Peace Treaty in San Francisco in 1952, pending the signing of a separate treaty settling issues between Japan and its former colony.[170] From the perspective of the Japanese government, however, the Koreans were identified as a particular group which was considered a source of social unrest or a welfare burden for Japan's economic recovery and comprised almost 90 per cent of all aliens in Japan.[171] When actual repatriation to North Korea began in 1959, the Japanese government had established formal diplomatic relations with neither South nor North Korea. Thus in the absence of any alliance between the host state and the divided homelands, an exclusionary policy was conducted (**Configuration 3**). It was, however, actualized thorough repatriation agreements between the Red Cross Societies of both Japan and North Korea in order to give the impression that the process was apolitical and had a purely humanitarian purpose. The key point here is that the main factors underlying the exclusionary policies of host state elites are not only interstate relations between the host state and the external homeland but also foreign policy preferences, such as revisionism. This is consistent with Mylonas' nation building theory. Most of the Japanese leaders who served from 1945 until 1964 acknowledged that Japanese colonization policies had completely failed to construct a national consciousness amongst Zainichi Koreans as being 'Japanese.' Immediately after World War II, Japanese Prime Minister Yoshida's policy of choice was repatriation or even forceful deportation, a policy which could not be fully effected during his administration. It went ahead after the transition of power in Tokyo from Yoshida to Hatoyama, by which time the Japanese elites' foreign policy had gradually shifted from preserving the status quo towards a more revisionist outlook.

After the US–Japan Mutual Security Treaty had been renewed in 1960, the US pursued its own strategy in East Asia and expected Japan to play an important back-up role in achieving this. Since rapprochement between Japan and South Korea was a core aim of US policy, the US pushed Japan to curtail the repatriation programme as soon as the treaty had been renewed by offering an alternative solution for the Japanese government's domestic problems; that a 'democratic Korea with settlement funds provided by Japan might appear more attractive to the Korean residents than repatriation to North Korea.'[172] Indeed, the strength of the US–Japan security alliance not only served the security of Japan but also allowed the US to compel Japan to accept the Republic of Korea as its principal Asian political ally against the communist bloc. The fate of the Zainichi Korean diaspora would inevitably be affected by what Victor Cha called the 'Quasi Alliance' between the US, Japan and the Republic of Korea.

4

Alliance Cohesion Matters: Japan's Policy towards the Zainichi during the Cold War Era (1965–80s)

Introduction

As the enthusiasm for repatriating the Zainichi to North Korea waned in the mid-1960s, it became increasingly clear that those who had settled in Japan were planning to remain permanently. Since the process of nation-building is one through which governing elites strive to match the boundaries of the state and the nation,[1] Japanese elites were forced to confront a serious problem: how would the government integrate the massive population of pre-war migrants into the national people? The ambiguous identification of the post-colonial Koreans in Japan, however, shows that the policy of integration employed by the elite nation-builders did not develop in either a linear or a coherent form.

There are two broad approaches a nation can use to maintain or change its status quo – through *security* (the ability to maintain the current situation regarding issues that a nation wishes to preserve), or through *autonomy* (the degree to which it can pursue any desired changes).[2] But the motivation to form alliances changes in a system where alliances already exist, as a nation's capabilities change over time. In other words, a nation's autonomy and security will rise and fall depending on its own capabilities and the support it receives from its allies. Walt analyses the reason why some alliances persist by focusing on the role of hegemonic power, or a strong alliance leader. Alliance leaders can discourage the dissolution of alliances by bearing a disproportionate share of the costs of maintaining that alliance, or by offering material inducements to make a continuing alignment more attractive.[3] The alliance leader must also be significantly stronger than any of its potentially disloyal allies, in order to be able to bear the additional costs of enforcing (or persuading) compliance.[4]

Morrow disagrees, emphasizing the key factor of an enduring asymmetric alliances as a series of bargaining deals between security and autonomy with a view to retaining harmonious interests among weak and strong states.[5] Rather than pushing the additional burden-sharing of alliance leadership that might eventually erode the asymmetry of power on which that leadership depends, major and minor states reflect their different interests by bargaining between security and autonomy. Following such a pattern of asymmetric alliance, the US–Japan–ROK triangular alliance has endured for over half a century since the end of World War II as a classic asymmetric alliance. Focusing on the space generated by the shifting cohesion of asymmetric alliance, however, theories of alliance cohesion and the treatment of diasporic populations explore the shifting balance between security and autonomy which significantly affects the cohesion of the allied states, insofar as the space occupied by the diaspora moves, either by expanding or contracting in relation to the policies employed by the host state; the approaches of exclusion, assimilation or accommodation mentioned earlier. The causal nexus between alliance cohesion, diaspora and nation-building policies illuminates an important argument which states that a diaspora is likely to compromise the inequality between a weak state and a strong state in terms of rational calculations regarding the trade-off between autonomy and security. Although a stronger state has a greater range of choices in terms of foreign policy than a weaker state, playing the diaspora card can provide leverage for the weak state, which will make a choice to either accommodate or exclude the diaspora in order to promote its own national interests by making a rational choice between security and autonomy within the asymmetric alliance framework.

Japan and South Korea have high-growth economies which are vulnerable to disruption by exogenous political and economic shocks as well as changing geopolitical power configurations because of their high degree of dependence on the US and on American security guarantees.[6] Moments of uncertainty, when these stabilizing links have been called into question, have allowed Japanese policymakers to adapt to a changing power distribution within Northeast Asia, which inevitably exerts an influence on policy shifts towards diaspora groups. The presence of a diaspora is a useful tool for host states to exploit as an informal connection with the diaspora's homeland, regardless of non-alliance with an enemy country, and allows the host state to design policies to retain the group in the state on a case-by-case basis. Indeed, the theory of alliance cohesion and diaspora explores such ambiguous policymaking by the host state, stating that it goes beyond the binary logic of home-host state or enemy-allied relations and results in a contradictory identification of the diaspora.[7] In this chapter, by providing and examining empirical data, I will test the following two causal pathologies in the case of the Zainichi diaspora using the four time periods in the Cold War era (1965–80s) that are outlined in Figure 4.1.

Configuration 1: if alliance bloc cohesion is strong, the host state policy will be one of *exclusion* toward a diaspora group supported by an enemy state, but one of *inclusion* toward a diaspora group supported by an allied homeland.

Configuration 2: if alliance bloc cohesion is weak, the host state will *accommodate* a diaspora group supported by a homeland belonging to an enemy alliance bloc in order to expand its own autonomy.

The chapter will then analyse whether these hypotheses would be supported by empirical explanations of the policy shifts on the Zainichi diaspora group. The policy shifts implemented by Japan in response to the exogenous political and economic shocks under asymmetric relations within the US–Japan–ROK alliance will demonstrate how Japanese nation-building process during the Cold War affects the implications of my own theories on the subject.

Period	External politico-economic shift	Alliance cohesion	Predicted policy towards the Zainichi diaspora
1965–70	Japan–ROK Normalization Treaty & Nixon Doctrine	**strong** alliance	**Configuration 1 inclusive** only to common **allied** diaspora
1971–4	Nixon Shock Japan–China Normalization Treaty	**weak** alliance	**Configuration 2 accommodation** to **enemy** allied diaspora
1975–9	US withdrawal from Vietnam & US–China Normalization Treaty	**strong** alliance	**Configuration 1 exclusion** to **enemy** allied diaspora
1980s	Plaza Accord & sharp revaluation of the Japanese yen against the dollar. The difficult relationship between the ROK and Japan over the textbook controversy	**weak** alliance	**Configuration 2 accommodation** to **enemy** allied diaspora

Figure 4.1 Predicted policy-shifts toward the Zainichi during the Cold War era

Empirics: Alliance Cohesion as a Causal Factor in Japan's Policy towards the Zainichi Diaspora in the Cold War Era (1965–80s)

Strong Alliance Cohesion: A Time of Cooperation (1965–70)

Japan–ROK Normalization Treaty: A Bargaining Deal between Security and Autonomy among Asymmetric Allied States

Although Japan and South Korea have never been parties to any bilateral defence treaty, they do exhibit *de facto* security ties as a result of common alliances with the United States.[8] Ongoing bargaining deals between security and autonomy have facilitated the Japan–US–ROK triangular alliance, helping retain harmonious interests between weak and strong states. Such harmonious interests between the United States, Japan and South Korea flourished from 1965 to 1970 under an asymmetric alliance.

Differences in capabilities are often highlighted by asymmetric alliances, and evaluating the trade-off between autonomy and security for a particular nation will depend on how the member countries of that alliance separate various issues under the headings of autonomy and security.[9] The primary factor for setting up the US–Japan–ROK triangular alliance was the erosion of American influence in the Western Pacific, which caused particular concern to United States policy makers during this period. The United States share of world export trade decreased from 16.5 per cent in 1955 to 15.9 per cent in 1960. During the same period the combined export share of France and West Germany increased from 11.8 per cent to 15 per cent. In contrast, by 1965 Japan's Gross National Product (GNP) had become the fourth highest in the world, behind only the USA, the USSR and West Germany.[10] On the international front, Japan and Germany's postwar economic growth presaged an increased role in world affairs that was growing as quickly as their capabilities.[11] Thus it was vital for the USA to establish a friendly and cooperative relationship between Japan and South Korea, and the two countries were encouraged them to seek an early conclusion to the negotiation of the Japan–ROK Treaty in order to lighten the US economic and defence burden in Asia.[12] After fourteen years of extremely difficult negotiations, the governments of South Korea and Japan signed the seven-article Treaty on Basic Relations on 22 June 1965 as well as other agreements, protocols, notes on property claims and economic cooperation, fisheries and cultural assets. In terms of cost benefit in the bargaining calculations, the Japan–South Korea diplomatic normalization contributed significantly to a reduction of South Korean economic dependence on the US.

For South Korea, the most serious problem was how to raise funds and maintain continuous economic growth under a gradual cessation of US economic and military aid. In 1965, Japan's per capita annual income had reached $682, whilst that of South Korea was estimated at little over $100. The agreement regarding Korean claims against Japan and on economic cooperation obliged Japan to provide $300 million in goods and services to be supplied to Korea in equal shipments spread over 10 years, as well as $200 million in repayable grants and $300 million in credits on a commercial basis.[13]

Needless to say, the United States was pleased with the successful outcome of its patient, friendly, and cooperative years of diplomacy in cementing this pivotal link in the anti-Communist strategic arc in East

Asia.[14] Diplomatic normalization with Japan was sought by both the Chang Myon government and the government of President Park Chung-Hee as a means of expanding Korea's foreign influence beyond its relationship with the United States. Since the late 1960s, there has been a *de facto* division of labour between the United States and Japan insofar as while the US defence commitment to South Korea has remained intact, Japan's economic relations with South Korea have steadily expanded.[15] By 1967, Japan had surpassed the United States as South Korea's primary trading partner, a position which Japan has maintained ever since.[16] Indeed, Korean–Japanese economic relations have grown enormously since normalization in 1965, and have brought extremely rapid economic growth to South Korea.

But what about Japan's interests? To Japan, the South Korean market was not important at all. The total budget of the South Korean government in 1965 was smaller than six months' sales figures for the Yawata Steel Production Company.[17] Japan's primary interest was not economic, but centred on the expansion of its own autonomy. Since Eisaku Sato became Prime Minister in November 1964, his priority issue was the US return to Japanese sovereignty of the Ryukyu Islands, which had remained under US administration since the occupation. His older brother, former Prime Minister Kishi, was intensely suspicious of the Chinese communists with their Confucian-Marxist view of the world, and Sato also believed that Japan's recovery as a great power could best be realized through continued cooperation with the United States and its Asian–Pacific allies.[18] The previous Prime Minister, Hayato Ikeda, had employed a two-Koreas policy by emphasizing the fact that the Japanese government did not and would not ignore the existence of the Northern regime, and that some cultural and economic relations with North Korea should be maintained. This policy was called *seikei bunri* – a policy which separated politics from trade and sought to build bridges with Japan's communist neighbours (especially Beijing) and work towards making space for adjusting policies in accordance with changes in the international environment. The policy aimed at preserving Japanese security by promoting relations with the two Koreas, thereby maintaining a balance of power on the peninsula.[19] The majority of public opinion among organized labour, intellectuals and students shared the same two-Korea approach, arguing that capitalism was the reason for imperialism and that imperialism was the source of world tensions. In their opinion the United States, as the leading capitalist country, was the chief threat to peace, while the socialist countries – Communist China and the Soviet Union – were working towards peace. Such thinking naturally led to a desire to break the US–Japan security treaty and achieve a strictly neutral position, exemplified by a two-Korea position for Japan.[20] In addition, many Japanese claimed that by solidifying the existing dividing line between North and South, the normalization treaties would obstruct the possible reunification of Korea.

Given that Japan's domestic situation was becoming one of increased political tension between neutrality and alignment with the United States, Sato's motivation to pursue the normalization of relations with South Korea went far beyond simple economic ambition. Even at the cost of its own security, the recovery of its own territorial sovereignty (the reversion of Okinawa) was Sato's primary policy, and one that would symbolize for Japan an end to the painful memory of World War II.[21] To be sure, brokering a normalization deal between minor states such as Japan and South Korea, and the reversion of Okinawa, would be acceptable to the US as a major partner, as long as the vital factor of maintaining a harmonious status quo was guaranteed by their continuing asymmetric alliance. Security was an indispensable factor for maintaining a stable status quo in order to preserve peace and stability in Asia. In the late 1960s, the United States was prepared to accede to intense nationalistic feeling in Japan and give the Ryukyu Islands back, but in turn it required that Japan grant the United States the unrestricted rights to use American bases in Okinawa for the defence of Korea, Taiwan and Vietnam. President Richard Nixon set out the Guam Doctrine in July 1969, calling for a link between Japanese and South Korean security.[22] In response, the so-called 'Korea clause' was delineated by Prime Minister Sato at the National Press Club speech in Washington on 21 November 1969. Sato said:

> In particular, if an armed attack against the Republic of Korea were to occur, the security of Japan would be seriously affected. Therefore, should an occasion arise for the United States forces in such an eventuality to use facilities and areas within Japan as bases for military combat operations to meet the armed attack, the policy of the government of Japan towards prior consultation would be to decide its position positively and promptly on the basis of the foregoing recognition.[23]

The clause was a culmination of prolonged Japanese–American negotiations on the reversion of Okinawa to Japan.[24] The US–Japanese joint communiqué and the speech by Prime Minister Sato were significant to South Korea, since the South Korean government was very much concerned that the reversion of Okinawa to Japan would result in the closure of the military base on Okinawa, which would threaten the security of South Korea. The 'Korea clause' and the Okinawan base agreement signified the shift of Japan's own position from neutrality towards both Koreas into alliance with the allied partner, South Korea, simultaneously highlighting the closeness of the security ties between Japan and the ROK under an asymmetric alliance.

The Impact of the Normalization Treaty on the Zainichi

The closer relationship between Japan and South Korea after the conclusion of the Normalization Treaty was not based on the mutual understanding of the people of either country. Many Japanese were afraid that Japan would be forced to involve itself in conflicts on the Korean Peninsula.[25] Some Koreans worried about the basis on which South Korean security would be maintained. The Normalization Treaty was signed in the midst of student demonstrations in Korea and leftist opposition in Japan. Complicated problems also arose in terms of the settlement of pre-war colonial migrants in Japan, especially during the process of negotiating the Normalization Treaty, due to Japan's policy shift toward the Korean peninsula and the way it was reflected in the treatment of Korean nationals in Japan. Article 3 of the 1965 Treaty on Basic Relations between Japan and the Republic of Korea recognized the South Korean government as the only lawful government in Korea, as specified in Resolution 195 of the United Nations General Assembly.[26] The treaty was accompanied by an agreement on the legal status of South Korean nationals in Japan, under which permanent residence would be granted to Koreans who had been resident continuously in Japan since 15 August 1945 and those who were born in Japan after 16 August 1945. Application for this status would be closed on 15 January 1971, but children born after 16 January 1971 to parents who had been accorded permanent residence would automatically be granted such status as well.[27] The Zainichi Koreans had to accept that they were South Korean nationals individually in order to receive permanent residence status, which meant that pro-North Korean members of the diaspora did not qualify.

The treaty, however, was not welcomed by many Koreans in Japan, for several reasons. First, the citizenship provided by South Korea had little substantive meaning in terms of the possession of rights and duties. By the mid-1960s, the majority of the Zainichi diaspora were second generation and Japanese was their first language, apart from some who attended ethnic schools and who could speak fluent Korean. By the conclusion of the treaty, less than 20,000 Koreans had applied for permanent residence, even though the Japanese authorities had expected to receive at least 5,000 applications a month.[28] As Thomas Hammar argues, all citizenship has a meaning that is both formal and substantive. Formally, it is understood as the membership of a state, and substantively it represents the possession of a number of rights in and duties to that state.[29] For example, the state of South Korea made a guarantee to the Zainichi Koreans that those who applied for the new status would be exempted from mandatory military service, but national voting rights in South Korea for citizens abroad were not granted until 2012.[30] Accordingly, South Korean citizenship was considered little more than symbolic, a clear verification of the bearer's alignment with South Korea and his or her rejection of North Korea. The policy was aimed at producing as many Japanese Koreans as possible who possessed a South Korean passport, rather than welcoming the diaspora as a whole to Japan's own co-ethnics.

Secondly, as far as the majority of the Zainichi Koreans were concerned, South Korean President Park Chung Hee was not popular at the time. According to the *Japan Times* in 1963, of the 643,000 Koreans in Japan, 375,124 were registered as North Koreans, 194,051 were registered as South Koreans and the rest remained neutral.[31] Although Park spent most of the 1960s as a civilian president, his approach to rule throughout the decade was militaristic, from the way he gained power through a coup to the regimentation of politics, society, and the economy that his reign implemented. To maintain surveillance and control over unsavoury elements and opposition political figures, the notorious Korean Central Intelligence Agency (KCIA), the internal security apparatus of the military government, was established in 1961.[32] As such, most of the Zainichi diaspora labelled Park's rule as a military dictatorship. For the Zainichi, the normalization treaty simply tied Japan to a closer relationship with the Park Chung-Hee dictatorship in South Korea. This was unsatisfactory for them because it involved the renunciation of Korean claims to compensation for colonial wrongs in exchange of a mass injection of funds into the Park regime's ambitious industrial development projects.[33] Prime Minister Sato gave a personal commitment to Seoul that he would attempt to persuade Japanese Koreans to move their allegiance from North to South.[34] In fact, the Japanese government actually kept this promise not by changing the Zainichi's allegiance to Pyongyang but by other means; the status change was accepted only from *Chosen* to *Kankoku*, 'South Korea'. In other words, once Zainichi Koreans had been registered as South Korean nationals, they would not be able to return to their previous status simply as Koreans.

This process of registration, however, created a problem which at first sight seemed rather trivial, but which in practice was to have an impact on the destiny of Japan's Korean community. When the word *Chosen* appeared in the nationality records of Japanese Koreans (the term the Japanese government and the occupation authorities used after members of the diaspora were stripped of their Japanese nationality by the Japanese government) the word was no longer recognized as the name of the Korean state. However, when the application for the status change of Zainichi Koreans into South Korean nationals went ahead, the nationality was also changed from *Chosen* to *Kankoku*. The implication of this was that *Chosen* related only to citizens of the DPRK, even though neither term, *Kankoku* or *Chosen*, had any legal substance as far as nationality was concerned. The registration completely excluded those who remained neutral and rejected the choice of either South or North Korea by opposing the division of their country. The Workers' Party of North Korea denounced the treaty as the selling-off of the nation, stating that the treaty had been engineered by US imperialism.[35]

After normalizing its relations with South Korea, Japan became increasingly less enthusiastic about trading with North Korea. The Sato government also refused to grant visas to North Korean technicians who wanted to come to Japan for technological training.[36] Inevitably, it became clear that Zainichi who registered as South Korean nationals hardly met with their family in North Korea during that time. The genuine agenda of recovering its essence as a Korean nation or nation state was gradually becoming estranged from both the Korean diaspora in Japan and the divided homelands. Indeed, the alliances between Japan, the US and the Park regime had a dual purpose – as a propaganda exercise that had nothing to do with the feelings of the diaspora and as a gradual marginalization of North Korea. The threat of deportation and other pressures from the immigration authorities were eliminated for those who registered as South Korean nationals but not for the pro-North Korean diaspora. But what concerned those who chose to be part of the pro-North Korean diaspora in the aftermath of the 1965 normalization treaty was that their enemies (Japan) would send them to South Korea, where they would be imprisoned.[37] Ryang argues that threats, such as deportation to South Korea, or discourse concerning the threats worked brilliantly as a *raison d'être* for *Chongryun*, and thus served to further reinforce the growing division within the ethnic community by reflecting the complete division of the homeland as a result of Cold War geopolitics. For example, when the Campaign to Visit Homeland was launched in the 1970s by *Mindan* with the backing of the South Korean government, *Chongryun* Koreans warned their members:

> The enemies use callous tricks. They take advantage of our memory and sentiment. Who does not want to visit the homeland? I believe my cousins are still alive in my native village [in South Korea]. But whenever enemies sent me brochures, I told myself that I should visit my village only after the fatherland's reunification under the wise guidance of the Great Leader. We must always watch out. Otherwise the enemies would try to erode our organization from within.[38]

Indeed, both external and internal factors play a key role in the way competing Korean identities have developed within the diaspora in Japan, particularly in the post 1965 era. The years 1965–70 provide evidence that the strong cohesion of the US–Japan–ROK alliance produced a harmony of interests between all parties by bargaining between security and autonomy. But at the same time it supports the argument that host state policies will only accommodate a common allied diaspora group under conditions of strong alliance cohesion (**Configuration 1**). Only those who registered as allied nationals received permanent resident status that offered rights and benefits in Japan that had been previously closed to them. This was in sharp contrast to those who did not register as South Korean nationals, the pro-North Korean diaspora.

The significant factor here is that the legal status of Koreans in Japan was determined by an international treaty, even though Japan as a sovereign nation had exclusive jurisdiction over aliens.[39] Domestic factors had little impact on the issue. In fact, Japanese public opinion did not support the treaty, particularly concerning the legal status of the pre-colonial Korean residents in Japan. Major Japanese newspapers pointed out that permanent residential status granted to Zainichi Koreans – who comprised 85–90 per cent of all alien residents in Japan – might pave the way for minority problems in the future.[40] What if those Koreans who became eligible to reside in Japan permanently, allowing themselves entitlement to many welfare benefits, did not want to naturalize themselves forever?[41] The Japanese government had been reluctant to grant permanent residence to successive generations of Japanese Koreans, so the so-called treaty-based permanent residency applied only as far as the third generation of Koreans in Japan, with the stipulation that the matter would be revised in 1991 to determine the status of future generations of Korean residents.[42] It was clearly a concern that the privileged alien status would be perpetuated forever. Legal discrimination between South and North Koreans within the same ethnic population might increase further ethnic antipathy or embroil Japan in an unnecessary international conflict.[43] However, from the Japanese perspective, it was more important that the US return the Ryukyu Islands to Japanese sovereignty. The islands had remained under US administration since the occupation, and in order to increase its own autonomy regarding the sovereignty of Okinawa, Japan accepted the so-called 'Korea clause', demonstrating to the US a change in Japan's own position from neutrality towards both Koreas to alliance with the US-allied partner, South Korea.

The South Korean government, however, treated Koreans from Japan as foreigners and criticized their inability to speak the Korean language rather than welcoming them as their own citizens. Even at the negotiation table with Japanese government, South Korea either made no serious effort or was incapable of exerting its influence to obtain better legal status for its nationals in Japan.[44] The ROK's main concern was to press for South Korean nationality for the Zainichi as strongly as possible – but only in terms of scoring points against the DPRK. The agreement therefore also included the goal of reducing pro-North Korean diaspora in terms of both numbers and influence by enabling the Japanese government to deport any Korean to South Korea, including those who claimed their nationality to be North Korean, because the South Korean government was obligated by the agreement to accept all Korean deportees from Japan. For both Japan and South Korea, it was important to eliminate the pro-North Korean diaspora from Japan as much as possible in the context of the US Cold War strategy in Northeast Asia.

It can also be seen that the cooperative relationship between Japan and South Korea during the period from 1965 to 1970 was not based on domestic factors or any kind of national consensus, but rather on external geopolitical factors. For minor states to protect each other against a major threat, Japan and South Korea needed to unite for the pursuit of peace and prosperity regardless of each state's own national consensus. But without creating a harmonious national consensus, an interpermeable political system would not be possible, and numerous problems between Japan and South Korea remained unresolved. No agreement was reached on the question of sovereignty of the Takeshima (Dokto) Island, and the longer term problems of resolving residence rights of subsequent-generation Koreans in Japan was postponed for twenty-five years (until 1991), leaving unresolved issues on the status of Koreans in Japan.[45] Under these conditions, it is easy to predict that a cooperative relationship could not be maintained in the long-term.

Weak Alliance Cohesion: Détente and Conflict (1971–4)

Expanding Autonomy under Weak Alliance Cohesion: 'Nixon-Shock' and the Japan–China Normalization Treaty

Stephen Walt considered the shared threat perception to be the most important deciding factor for alliance solidarity between states.[46] Walt's alliance theory explains how US decision makers conducted diplomatic normalization with China and simultaneously pursued a reductionist policy with Moscow, exploiting Russians concern about being isolated from the détente procedure.[47] Nixon's announcement of his plan to visit China was formulated in a broader context to maintain a favourable balance of power, considering the changes in East Asia and the world as well as its own military capability. By carefully managing the triangular relationship between the US, China and the Soviets, the US was trying to enhance American control whilst simultaneously achieving a more stable balance in international politics.

On 14 July 1971, President Richard Nixon surprised the world with a televised announcement of his intention to visit China. The subsequent Shanghai communiqué confirmed the two states' agreement on the principles of peaceful coexistence and anti-hegemony in Asia, including a mutual aspiration for normalized diplomatic relations.[48] As a result, the Chinese ended the legacy of the Korean War in their relations with the United States, breaking down the traditional American policies of diplomatic isolation, military containment and economic sanctions against a stable China.[49] The underlying logic, however, was to increase US leverage in world politics while avoiding the rekindling of a Sino–Soviet alliance. With Beijing and Washington engaged in a dialogue, Nixon and his special assistant Henry Kissinger reasoned that the Soviets would have to work harder to sustain and improve relations with Washington in order to keep China and the United States from becoming too close.[50]

In the early stages, this strategy was very effective, reflecting the well-known Chinese proverb; 'killing two tigers with one stub.' For the United States, occupying the central position in a US–China–Soviet Union strategic triangle reduced the possibility of wars on two fronts, in Asia and Europe. The Sino–American relationship not only reduced the potential for conflict in Asia but also affected US–Soviet relations. In October 1972, Washington and Moscow signed a comprehensive trade agreement linking to a settlement of World War II land-lease aid and the restoration of most-favoured-nation status for Soviet exports. It also included promises of US agricultural credits, allowing the Soviets to purchase hundreds of millions of dollars' worth of US grain. Most importantly, both nations reached agreement on the Strategic Arms Limitation Talks (SALT), the Anti-Ballistic Missile (ABM) treaty and the interim agreement on offensive arms limitations.[51]

In this case, however, as Walt argues, the US was making an attempt to balance power. Nations will choose to balance power for two reasons. Firstly, a state may choose to ally itself with a weaker state rather than a dominant power by joining with those who cannot readily dominate their allies in order to avoid being dominated by those who can. Secondly, joining the weaker side increases the new member's influence, because the weaker side has a greater need for assistance. In the same way, Nixon and Kissinger advocated rapprochement with China rather than the Soviet Union because in a triangular relationship, it was better to align with the weaker side.[52] In his memoirs, Kissinger argued that he initiated a special relationship with Beijing, supplying sensitive intelligence information, including satellite photography on various occasions. In this regard, Kissinger was treating the Chinese, but not the Soviets, as well as he treated the United States' NATO allies.[53] By exploiting their anti-Soviet animus, the US sought leverage with a Chinese leadership which totally disagreed with intrinsic US values as to where the world was going, because the US felt it could do nothing against China, a country which, through its massive population, might grow to such proportions as to make it hard to deal with in 20–50 years' time.[54]

So why did the Soviet and the Chinese leaders arrive at the decision to meet with US leaders? The Soviets would not have been interested in arms control or cooperation were it not for internal economic problems and the possibility that China might pose a threat to the East. By the same token, Mao would not be talking to the leaders of the capitalist world and courting the US unless he was concerned by the Soviets.[55] Walt explains that threatened states face a choice between balancing (allying against) and bandwagoning (allying with) the threatening power. The choice between conciliation and balance involves optimizing security, autonomy and intrinsic national values: conciliation buys security at some cost to intrinsic values; balancing buys it at the cost of autonomy.[56] In the early stage of their negotiations with the US, China made a choice to buy security since it feared the Soviet Union – especially in the face of the build-up of a million Soviet troops along its borders during the period of Sino–Soviet conflict. However, conciliation or a cooperative moves toward an adversary may weaken an ally's cohesion or even trigger an ally's defection. The greater the threat posed by an adversary, the greater the cohesion of the alliance will be. Morrow explains that the balance of power calculation among the major states ignores the role of minor powers. In fact, alliance and adversary relations are a seamless web; any change in one will impinge on the other.[57]

Indeed, when Nixon announced his visit to China, the Soviets not only saw the approach to China as being aimed against them, but were also worried about the impact of the US move on Japan. The Soviets had already suggested that Nixon's shock moves – the ten per cent surcharge on imports and the failure to provide Tokyo with due notice of the administration's *volte face* on China – could prompt the Japanese to move closer to Beijing.[58] Nixon and Kissinger were worried that European countries might retaliate against the United States due to their new economic policy decisions, and arranged meetings in December with European heads of state to help forge a consensus on international economic policy.[59] However, the US failed to recognize the impact of its policy change on its main ally in Asia, Japan. The new US policy was announced on 15 August 1971 in a national televised address. Nixon declared that the window on gold would be closed and that gold would no longer be transferable into US dollars in order to control inflation and to correct a gaping trade deficit and a weakening dollar, and that the US would impose a 10 per cent import surcharge. Moscow's prediction was right in terms of Japan's response to the so-called 'Nixon-shock'. It allowed Japan to increase its own security and simultaneously to expand its autonomy without having to make a trade-off between the two under an asymmetric alliance. Japan secured its safety by gaining better relations with China and the Soviet Union and was able to enjoy new economic opportunities with the communist powers, yet still had the luxury of remaining within the shadow of the US security umbrella after China dropped its explicit opposition to the US–Japan defence treaty.[60]

Given the key issues surrounding Japan's contentious political debate between neutralism and alignment with the United States since the end of World War II, one of the most serious and long-standing strains on the Japanese–American relationship had resulted from popular feelings in Japan of regional cooperation. Because of its geographical proximity to China and its long and close cultural association and mutual economic interests, the Sino-American détente inevitably allowed Japan to seek fuller and friendlier ties with Communist China much more easily than the Americans could. The Japanese opposition parties, for twenty years the most persistent advocates of rapprochement with China, also found themselves jostled aside in the Conservative stampede.[61] The process of rapprochement between Japan and China culminated with the September 1972 Tanaka-Zhou summit. The joint communiqué issued at the conclusion of the summit affirmed the determination of both states to normalize relations and to uphold the principles of peaceful coexistence and mutual respect for territorial integrity. The key factor in reaching the rapprochement so quickly was the Tanaka government's decision to scrap the all-important Taiwan clause, which had been opposed by Beijing as the primary impediment to amicable relations. Prime Minister Tanaka, unlike Sato, accepted the government of the People's Republic as the sole legitimate government of China and recognized Taiwan as part of its territory.[62]

The One-China policy automatically caused a reinterpretation of the 1969 so-called 'Korea clause' by Japan. In other words, it was no longer the security of the ROK, but that of the entire Korean peninsula that suddenly became essential to Japan. Also the permission by the Japanese for its bases to be used for ROK defence should no longer be presumed as automatic. Japan's reinterpretation of the Korea clause and the Okinawan base agreement implied a shift in Japan's position towards the Korean peninsula to one of neutrality, a move that eroded the cohesion of US–Japan–ROK security ties under asymmetric alliance. The US wanted to reaffirm Japan's commitment to the original Korea clause adopted in 1969, as America wished to see Japan offer maximum political support to the South Korean government and increase its indirect assistance to South Korea to help it maintain parity with North Korea.[63] From the Japanese perspective, however, the original Korea clause of 1969 had become an outdated policy, and the Japanese wanted to move towards a New Korea clause that would underscore the idea that maintaining peace and security for the whole Korean peninsula is essential for Japanese security.[64] In an asymmetric alliance, only South Korea – one of the minor states – would be unable to improve its position regarding either security cost or autonomy. Even though President Park expressed willingness in January 1971 to seek rapprochement with the socialist countries he characterized as non-hostile, such as the Soviet Union and communist China, China maintained a rigid one-Korea policy and rebuffed Seoul's overtures.[65]

Given the situation of Sino–Soviet conflict among allied states, North Korea became strategically far more important. It was indispensable to China and Moscow, but Pyongyang could not increase its power in the triangular relationship. However, as far as South Korea was concerned, the Sino–American détente did not provide any opportunity for a trade-off between security and autonomy. Park made a direct and confidential private appeal to Nixon, hand-delivered to Secretary of State William Rogers on 16 September. In the letter he urged the US not to accept Beijing's call for the removal of foreign troops from the Korean Peninsula.[66] In a response to Park's letter in a few months later, Nixon replied that as regards the withdrawal of additional United States troops now stationed in the Republic of Korea, it would be policy under the Nixon Doctrine not to reduce US forces overseas any more rapidly than would be consistent with the increasing capabilities of the host country.[67]

In such a situation, a state must increase its own capabilities to secure its own sovereignty. As a result, South Korean politics in the 1970s moved back to a type of brutal domestic repression that had not been seen since the crushing of the autumn 1946 uprisings. The Heavy and Chemical Industries plan and the *Yushin* (revitalizing) constitution allowed the kind of economic control that produced a powerful and worrying spurt of economic growth.[68] On 17 October 1972 Park instituted the *Yushin* reforms, which strove to expunge all political opposition and consolidate Park's hold on power by imposing martial law and dissolving the National Assembly. Opposition politicians and dissidents were imprisoned and tortured and society in general was subject to extreme repression until Park's death in 1979.[69]

The Nixon administration publicly distanced itself from Park's undemocratic actions and pointed out that Park had held no prior consultation with the US before instigating his repressive policies. When South Korean Foreign Minister Kim Yong Shik asked US ambassador Philip Habib to issue a statement in support of Park's decisions, the United States refused to do so, eventually adopting a policy of non-interference in South Korea's domestic affairs.[70] This was no doubt because the conciliation of an adversary state's cooperative moves towards another adversary would weaken the cohesion of the main alliance. Subsequently, North Korea began to cultivate informal contacts with Seoul's two principal allies – the United States and Japan. Kim Il Sung asserted a degree of autonomy in foreign affairs and attempted to turn the Sino–Soviet competition to his advantage using an alternative view of the asymmetric alliance. He initiated his country's diplomacy toward the United States in the early 1970s and expressed his willingness to establish diplomatic relations with Japan without the abrogation of the Japan–South Korea treaty for diplomatic normalization,[71] which inevitably provoked a further policy shift toward Japan's treatment of Koreans in Japan who had been excluded during the normalization treaty between Japan and South Korea.

Among opposition party politicians, one of the more influential leaders who supported the initiation of diplomatic relations with North Korea was Tokyo Governor Ryokichi Minobe. Prior to détente, the Japanese government in general discouraged visits by ruling or opposition party politicians to North Korea, as these were seen as unnecessarily troublesome for Japan–ROK relations. It was Tokyo Governor Minobe who, with the government's approval, visited Pyongyang in October 1971 for the first formal meeting with Kim Il Sung. Following Minobe's trip, in November 1971, 246 Japanese legislators formed the Diet's League for the Promotion of Japanese–North Korean Friendship. This body called for the promotion of Japanese–North Korean trade, and diplomatic relations between the countries to increase bilateral ties and eventually negotiate a normalization treaty.[72] In 1973, long-term low-interest loans from the government-sponsored Export-Import Bank to North Korea were approved, allowing Japanese exporters to use their funds for transactions worth 500 million yen. The increase in trade also brought numerous North Koreans into Japan. A year before the approval of the Export–Import Bank application to North Korea, in 1972, a seven-member delegation from the DPRK Committee for the Promotion of International Trade visited Japan for the first time, followed in 1973 by some half-dozen delegations of North Koreans involved in radio and television broadcasting, steel, iron, cement, and other industries.[73]

Accommodation Policy: Diaspora as a Useful Tool for Expanding Autonomy during Détente

The space of autonomy created as a response to détente allowed the Japanese policy makers to view the pro-North Korean diaspora as a useful group for mediating various exchanges between Japan and North Korea, and supported a policy of accommodation to make them more flexible in terms of crossing state boundaries and thus increasing potential capabilities. In July 1972, under the leadership of Governor Minobe, the Tokyo Metropolitan

Government provided a tax-exempt status for the pro-Pyongyang group by saying it recognized that their properties were being used for a state delegation in Tokyo and as such were exempt from taxation under the Vienna Convention.[74] Since then most of the *Chongryun* facilities have been either entirely or partially exempted from fixed-asset taxation. Because there is no formal diplomatic relations with Pyongyang, the *Chongryun* headquarters in Chiyoda Ward is generally considered North Korea's *de facto* embassy, ever since the pro-Pyongyang group began building on properties in Japan in the mid-1960s. From the legal point of view, however, the *Chongryun* cannot be seen as North Korea's administrative organization but can only be considered as a voluntary group. As a matter of fact, the *Chongryun* did precede the issue of visas for Pyongyang as part of the North Korean repatriation project. Thus the tax-exempt status for the *Chongryun* was deemed to be formal recognition of their status as an actual North Korean administration organization, regardless of diplomatic relations.

Another accommodation policy toward the pro-North Korean diaspora group was the issue of re-entry visas. During the 1971–4 period, Japanese immigration regulations showed greater flexibility in visa policies, beyond the traditional criteria of granting visas only on humanitarian grounds. In March and August 1972, the justice ministry approved re-entry visas for a number of prominent *Chongryun* leaders with full knowledge that their visits would be for political reasons – to attend the celebrations in Kim Il Sung's birthday – rather than filial purposes.[75] The ministry approved re-entry visas for more North Korean residents in September 1971 than the total granted for all of 1969 and 1970.[76] This action showed that the Japanese justice ministry was expanding the categories for which visa applications would be accepted due to government's expectations for increasing Japanese–North Korean trade, cultural exchange and for building up bilateral ties and working towards an eventual normalization treaty.

South Korea strongly denounced Japan's new re-entry visa policy as a threat to national security, fearing that relaxed restrictions on travel for individuals to and from North Korea would provide an unrestricted space for destructive activities. Unfortunately these fears were well-founded. On 15 August 1974, a second-generation Japan-born Korean, Mun Se-Kwang, tried to assassinate the President on behalf of North Korea, but the gunshots killed his wife instead. The South Korean government claimed that

Mun had been trained by North Korean special agents linked to *Chongryun*, and that his intended target was President Park, whose regime he considered the main impediment to the unification of North and South Korea.[77] Foreign Minister Kim Dong Jo and Ambassador Kim Yong Son called for the outlawing of pro-North Korean organizations like the *Chongryun* in Japan. In addition, the National Assembly issued a resolution advocating the severing of diplomatic relations if Japan refused to assume responsibility for the incident.[78] The Tanaka government, however, resisted South Korean demands for an apology by saying that the assassination was a result of the repressive Yusin system rather than a North Korean-backed attempt to overthrow Park.[79]

As a matter of fact, the Mun Se-Kwang affair showed that the Japan–ROK normalization agreement on the status of Koreans in Japan had left many issues unresolved. Mun was a Korean resident in Osaka who had registered as a national of the Republic of Korea under the Treaty of Normalization with South Korea in 1965. Even for those who did manage to obtain permanent residence under the agreement, life in Japan was extremely difficult to survive in terms of employment, access to public funds and voting rights. Quiet and unofficial discrimination persisted in some private firms and even in relations with the ordinary Japanese people.[80] The 1965 Normalization Treaty implied that Zainichi Koreans would have three options: they could return to the fatherland, they could stay in Japan and become Japanese citizens, or they could remain as a permanent foreign community in Japan.[81] In practice, naturalization was a discretionary choice by the minister of justice and could be withheld if officials judged the applicant to be undesirable or insufficiently Japanese in terms of lifestyle.[82] All visible elements that connected to Korean ethnicities or roots had to be eliminated. For example, until 1986, anyone becoming a Japanese citizen was forced to take a Japanese-style name, and the naturalization process involved intrusive checks into the applicant's finances, family, friends and personal life.

As such, some of the Zainichi who were not sure if they should stay in Japan with an unclear future and an ambiguous identity – regardless of whether they were North or South in origin – supported the student movements opposing the dictatorship and pursuing the creation of an ethnically unified democracy in Korea which reflected a more positive solution to resolving their unclear identity. In order to find a way of escaping the moratorium situation, an *imagined* sense of the putative fatherland gradually brought what Benedict Anderson calls 'long distance nationalism'[83] to the younger generations of the Zainichi. The idea of democratization in Korea became a prevalent slogan used to capture the younger members of the Zainichi in the early 1970s. In Korea, the split between dictator and people was becoming clearer and the country was approaching an era of extreme politicization. In Japan the echoes reached the hearts and minds of the subsequent generations of Zainichi Koreans, resulting in an 'ethnic awakening.'[84] The increasing tendency towards an ethnic identity – or at least a consciousness of belonging to a specific ethnic group – allowed the Zainichi to mould a stronger sense of identity from a period of comparative ambiguity.

As Brubaker argues, the Korean diaspora – the Zainichi – should not be considered in substantive terms as a bounded entity, but rather as an idiom, a stance, a claim,[85] a factor which might one day evoke a spontaneous ethnic awakening. The years 1971–4 saw the majority of Zainichi, as Kang Sang-jung argues, becoming attached to a quasi-conceptual democracy by accepting a stance of 'ethnic nationalism = democracy = reunification',[86] and in these terms they aspired to reach the essentialization of belonging as an alternative space of territoriality in the name of co-nation.[87] In generalist terms, this approach and this growing ideology reflects the view expressed earlier by Brubaker, who stated that the diasporic concept may help redefine geo-political boundaries by stretching across the frontiers between states. In personal terms, it is obviously something close to what Mun Se-kwang felt when he set out on his mission to assassinate President Park. According to the *Yomiuri* newspaper, Mun expressed his thoughts in the form of a letter in which he said that he felt ashamed he had not written in Korean. The letter listed his previous experiences, such as discrimination in his school life, and also recalling his vivid impression of the beautiful blue sky when he arrived in his putative fatherland for the first time in his life, which was in such contrast to the aim of his visit, which was to assassinate President Park. In his thirty-two page letter, Mun expressed regret at what he had done, but also admitted that he could not eliminate the hope that he would have lived a different life to the one he had lived if such a thing were possible.[88]

Under the accommodation policy of the Japanese government towards the pro-North Korean diaspora since the early 1970s, *Chongryun* successfully constructed a pro-North Korean diasporic identity that served as the basis for mass mobilization. Since the 1970s, *Juche* philosophy has had a central place in all *Chongryun* cadre education programmes, and has ultimately played an indispensable role in resource mobilization. Juche literally means 'subject', although it is often translated as 'self-reliance'. According to Kim Il Sung, 'establishing *Juche* means that the people approach the revolution and construction in their own country as masters'. Japan's accommodation policy during the 1971–4 period virtually turned a blind eye to *Chongryun's* role as a supporter of Kim Il Sung. It was advantageous for Korean schools to remain free from interference. *Chongryun* school education openly involves the systematic training of individuals and the implementation of control through study meetings and other tactics.[89] The worship of Kim Il Sung and strict adherence to the rhetoric of his *Juche* ideology has remained central to all *Chongryun* cadre education programmes since the 1970s. In April 1972, to commemorate Kim Il Sung's sixtieth birthday, *Chongryun* waged a 'gift-sending campaign'. A total of 5,900 tons of gifts went to North Korea in nine shipments: the total expense amounted to ¥ 5 billion. Ideology, as Anthony Giddens conceptualizes, refers to ideological ideas being understood in terms of the capacity of dominant groups or classes to make their own sectional interests appear universal to others. Such a capacity for giving therefore represents a type of resource involved in domination.[90] By emphasizing *Chongryun* activities as being in the service of Kim Il Sung and the fatherland, *Chongryun's* officials could use them to appeal to their affiliates' patriotism while self-righteously denying any interest in personal advancement, claiming instead to work for the sake of the fatherland, the nation, and compatriots at large.[91]

With regards to domestic factors, Japanese media supported the Japanese government policy, which contributed to exacerbating Japan–ROK frictions during the 1971–4 period. Laudatory articles on North Korea, spurred by the proliferation of cultural and political exchanges, cultivated great curiosity and infatuation with the North within the Japanese public.[92] The undemocratic and unjust activities of the authoritarian regime led by Park, such as suppression of student strikes or civil protests against the government, allowed the Japanese media to adopt a more sympathetic attitude to North Korea whilst continuing with more negative coverage of Park regime, news of which became the subject of much conversation among Japanese citizens. While such variations in public opinion may well have allowed the Japanese government to adopt an accommodative posture towards the pro-North Korean diaspora more easily in comparison with Normalization Treaty in 1965, Japan's attitudes were driven more by the change of geostrategic threat perception. The sense of détente created by US foreign policy helped alleviate Japan's external-threat perceptions and allowed Tokyo to examine new diplomatic opportunities. Tokyo's visa policy towards the pro-North Korean diaspora and the tax-exempt status for *Chongryun* as formal recognition of its status represented a tactic by Japan laying the foundations for eventual normalization with North during the 1972–4 period.[93]

Nixon's visit to China was designed to ease tensions in Asia during the 1971–4 period, and would have effectively reduced its own economic and security burden in the asymmetric alliance. Unfortunately, due to its miscalculation of the other minor ally's behaviour, the US was unable to improve its position of leverage in world politics. Every time the US met with their Chinese counterparts and exchanged views on the question of normalization, Chinese leaders just repeated they would only accept the Japanese way.[94] Japan's quick move normalization with China ironically erased the effectiveness of the idea of the agreement – that Nixon did not challenge Beijing's 'one-China' principle – and fell into line with the idea that the US needed to maintain some form of defence commitment to Taipei. Contrary to the aim of reducing the threat, Japan's behaviour not only ruled out the possibility of détente but also increased the genuine concern that a breakdown in relations between Japan and South Korea was imminent. The period between 1965 and 1970 confirmed the hypotheses that the greater the threat posed by an adversary, the greater the degree of cohesion becomes. Alternatively, however, the years 1971–4 witnessed that conciliation of an adversary or a state's cooperative overture toward an adversary would weaken an ally's cohesion. The history of the period also supports the argument that if alliance bloc cohesion is weak, the host state (Japan) will accommodate a diaspora group supported by a homeland belonging to an enemy alliance bloc in order to expand its own autonomy (**Configuration 2**).

Strong Alliance Cohesion: The Carter Years and Burden-Sharing Period (1975–9)

Security Alerts and Strong Alliance Cohesion: US Withdrawal from Vietnam and the US–China Normalization Treaty

The task of American foreign and defence policy was too much inclined towards the global balancing of power, since balancing was the dominant tendency in international politics. Yet a major US build-up was hardly likely to spur its allies to greater efforts. Via the logic of balancing, they were far more likely to hitch a ride with the most powerful.[95] America's Asian allies had frequently been accused of flagrant examples of free-riding in both their economic and their security behaviour. Japan and South Korea had long been accused of protecting their domestic markets while demanding free access to markets elsewhere,[96] and the US military programme should be revised so that the unilateral US response could preserve its own commitments efficiently. Based on this logic, the Carter administration's disengagement plan from the Korean peninsula began in the wake of the fall of Saigon in the spring of 1975. Some US analysts such as Stephen Krasner and Robert Keohane viewed Japan as a nation that was availing itself of the benefits of the global free-trade regime and security guarantees provided by an American superpower without paying the costs.[97] In contrast, Calder suggests that Northeast Asian nations, particularly Japan and South Korea, played a substantial burden-sharing role in terms of Host Nation Support (HNS) for American forces in Northeast Asia, covering larger shares of local US military basing costs than virtually anywhere else in the world.[98] This burden-sharing not only set the background for the second period of cooperation from 1977 to 1979 in response to the Carter plan but also sparked a policy shift away from the dominant policies of the period between 1965 and 1970, a shift that was marked by the new Korea clause, political reconciliation between Tokyo and Seoul, a revival of economic ties and a uniform attitude toward North Korea.[99]

The American withdrawal from Vietnam in 1975 suddenly allowed the Japanese government to attend to its own security concerns rather than dedicating itself to expanding its autonomy. The US–Japan security treaty was limited geographically to the territories under Japanese administration, which meant that Japan was committed to some degree to cooperate with American policies in the general area of the Far East (Article VI). From the Japanese perspective, Vietnam was not included within the scope of the Far East. Vietnam was Japan's 'fire across the sea,' an attitude which allowed the Japanese to remain aloof.[100] The danger with the US retreat, however, was that the fire might now be re-focussed onto Japanese shores within the broader scope of the Far East.

To make matters worse, on 17 March 1976, Jimmy Carter, governor of Georgia and a leading contender for the Democratic Party's presidential nomination, claimed that it would be possible to withdraw US ground forces from South Korea on a phased basis over a specific time span to be determined after consultation with both South Korea and Japan. Disturbed by the Vietnam trauma, Carter felt that the United States should avoid any situation that might require its automatic involvement in another Asian ground war, and favoured instead a policy that would allow strategic flexibility and tactical mobility by giving higher priority to the Navy, Air Force, and Marines than to conventional ground forces. Carter regarded the planned military withdrawal as a tangible way of saving US taxpayers' money, which was consistent with his campaign themes of fiscal austerity and a balanced budget. However, what Carter failed to understand was the symbolic and psychological importance that South Korea attached to the continuing US military presence in Asia. In January 1977, even some of the US Joint Chiefs of Staff stood up in opposition to Carter's policy by saying that in Korea, American military presence was considered as a tangible manifestation of US commitment to the security of the Republic of Korea, and US presence helped deter North Korean aggression toward the South.[101] Disregarding the JCS recommendation, Carter declared that the United States would withdraw its ground troops from South Korea over a period of four to five years, but would leave behind adequate ground forces under South Korean command and would provide air cover for South Korea over a longer period of time.[102]

Ironically, the Carter plan to withdraw US troops from South Korea brought Japan and South Korea substantially closer to each other. In April 1975, Japan's Miki cabinet dispatched Foreign Minister Miyazawa to Washington, where he reaffirmed the validity of the 'Korea clause' in the Nixon–Sato communiqué. In meetings with ROK Premier Kim Jong Pil the following May, Prime Minister Takeo Miki stated that South Korea's security was essential to that of Japan and that Japan would no longer view Tanaka's 1974 reinterpretation of the clause as relevant.[103] Although relations between Seoul and Tokyo had severely deteriorated during the 1971–4 period, the two governments had been making efforts at resolving disputes since 1975. The May 1975 meetings between Miki and Kim Jong Pil produced an agreement to reinstate the annual joint ministerial conferences, suspended since 1973.[104] The shift in the Japanese attitude was also reflected in its policy on economic aid for South Korea. The eighth Japanese–Korean ministerial talks in September 1975 promised to provide public loans for various projects contained in South Korea's third economic developmental plan (1972–6) and the fourth five-year plan (1977–81).[105] Moreover, the most ground-breaking step in Japanese–Korean relations was that in April 1979 General Shigeto Nagano, chief of staff of the Japanese Land Self-Defence Forces, visited South Korea in his official capacity. In July, Ganri Yamashita, the director-general of the Japanese Defence Agency (effectively the Minister of Defence) paid an official visit to South Korea. Although these events had been in line with US Strategy policy – especially Dulles' long-term expectation for the America's Asian and Pacific allies – Carter's withdrawal plan from Asia finally spurred the Asian allies to greater efforts for Asian regional stability.

Carter's insistence on troop withdrawal could be seen as a sign of an overall retreat from the Pacific not only by Japan and Korea but also by other Asian nations. For the North Koreans particularly, the US military withdrawal from South Korea came as a welcome announcement. As a matter of fact, North Korea was at the time superior to its southern counterpart in every key aspect of military capability. China also reaffirmed its traditional support for North Korea, opposing any form of two-Korea solution and demanding the complete withdrawal of US forces from South Korea as well as the dissolution of the United Nations Command. In the Sino–North Korean joint communiqué of 1975, Chinese Foreign Minister Huang Hua specifically ruled out mutual recognition of North and South Korea and their simultaneous admission to the United Nations, declaring categorically that China did not recognize the South Korean authorities.[106] Even after normalizing diplomatic relations with the United States in December 1978, which confirmed the fact that China's détente policy toward the United States and Japan had been fully realized, China showed no intention of improving its relations with South Korea. In his meeting with Chinese vice-premier Deng Xiaoping in January 1979, Carter attempted to improve China's relations with South Korea and asked Deng to use his influence in North Korea to instigate inter-Korean talks. Deng rebuffed Carter, telling him that it was not yet possible to have trade relations or direct communications between China and South Korea.[107]

President Carter's plans for withdrawal, however, were short-lived. With the leak of new intelligence data concerning North Korean military capabilities in January 1979, which had allegedly ascertained that North Korea enjoyed clear ground superiority over South Korea, the Carter plan for withdrawing US troops from South Korea was doomed.[108] The signing of the Soviet–Vietnam treaty in November 1978 also gave the US reason to reconsider the military balance in Asia. On 20 July 1979, Presidential Secretary Advisor Zbigniew Brzenzinski announced Carter's decision firstly that withdrawals of combat elements of the Second Division would remain in abeyance, and secondly that while there would be a small reduction in US personnel, some support units – including one I-Hawk air defence battalion – would remain until the end of 1980. The timing and pace of withdrawals beyond these this date would be re-examined in 1981.[109] Carter's term in office, however, ended in January 1981.

Conventional alliance theory suggests that smaller minor powers have incentives to stop providing collective benefits such as national security before the optimal output for collective benefits of alliance can be fully realized.[110] Contrary to the expectations of collective action theory, Carter's neo-isolationist stance and a somewhat simplistic grasp of the complexities of the Korean situation paradoxically produced a strong cohesion between Japan and South Korea. Although Japan and the Republic of Korea used to be comparatively low providers of defence burden-sharing under an asymmetric alliance, Carter's unilateral announcement to withdraw ground forces from Korea made both minor states' defences assume extraordinarily high support levels. In May 1978, in the form of a 'sympathy budget', Japanese Defence Agency Director General Kanemaru proposed the expansion of HNS backing for US forces in Japan. On 23 August 1977, the Defence Facilities Administration Agency decided to cover ¥6 billion in local US labour costs, including half the wage costs and social welfare premiums for local employees.[111]

South Korea, although smaller in scale, had commonly been considered in a similar light. In 1977, following the fall of Saigon, the Korean government also began supporting the operational cost of the Joint US Military Assistance Group-Korea (JUSMAG-K).[112] In the mid-1970s, both Japan and South Korea made the decisive shift from the free-rider status predicted by alliance theory into a cohesive alliance under asymmetric alliance theories.

No More Accommodation Policy: The Boundaries Close, Opening a New Identity for the Zainichi

The new 'Korean Clause' essentializing South Korea's security to that of Japan inevitably forced Japanese policy makers to shift their attitudes towards the pro-North Korean diaspora, ending the policy of accommodation. During the 1975–9 period, the Miki and Fukuda governments substantially decreased their contact with North Korea. For the Japanese government, this meant that during that it became dangerous for the diaspora to cross boundaries or reach into enemy space, as this could be considered as playing a part in enhancing Pyongyang's war potential. Flexible movement became a thing of the past. When Kim Il Sung visited Beijing in April 1975, his statement clearly showed the North's willingness to capture this opportunity:

> If a revolution takes place in South Korea, we will strongly support the South Korean people as the same nation. If the enemy ignites a war, we will definitely answer with war. What we lose in this war is the military demarcation line but what we get is re-unification of our country, Korea. The key point is whether we get peace or war, depending on the US attitude. If US troops withdraw from South, we Korean nations will resolve our problem, re-unification by means of peaceful methods by ourselves.[113]

Given such a provocative statement, transnational diasporic activity became extremely dangerous. During the 1971–4 period, the Park government had demanded the disbanding of the pro-Pyongyang organization, *Chongryun*, for its involvement in subversive activities. The Tanaka administration, however, declined Seoul's request to regulate the organization, saying that the Japanese government would only involve itself if it violated

Japanese law.[114] However, during the period between 1975 and 1979, the Japanese government did acknowledge for the first time that *Chongryun* was a political organization tied to Pyongyang rather than an independent Korean body in Japan, and instituted subsequent restrictions to prevent transnational political activities.[115]

Among measures taken by the government was the denial of re-entry visas if visits to North Korea were associated with political purposes rather than social issues such as family reunions. In 1977, when the chair of *Chongryun* and a number of its prominent leaders applied for re-entry visa to attend the North Korean election as foreign delegations, their request was denied because it was obvious that the visits were for political rather than filial purposes. The Japanese justice ministry made it clear that the Japanese government had never issued re-entry visas for political reasons, but issued them only for visiting families and graves.[116] However, in March and August 1972 the Justice Ministry had approved re-entry visa for prominent *Chongryun* leaders – including the same people who were denied visas in 1977 – with full knowledge that visits were for political purposes.[117] The possible US retreat from Pacific Asia forced the Japanese government to view North Korea as an alarmist threat, and from that point on, Japanese policy makers viewed the diaspora group that supported Pyongyang as an extremely dangerous minority. The Japanese government for the first time publicly made its position clear, that *Chongryun* was a political group supported by the DPRK rather than a simple organization of Korean residents in Japan.

The cooperation between Seoul and Tokyo also became part of a more controversial argument with regard to surveillance of the resident South Korean nationals. Those who visited South Korea were subjected to strict government surveillance and some were arrested as spies working for North Korea. During the 1971–4 period, every time the KCIA sent 'summons letters' directly to Zainichi Koreans via the South Korean embassy without the prior consent of the Japanese government, the Justice Ministry gave the South Korean government a warning by saying that this act was a clear infringement of Japanese sovereignty.[118] The KCIA was, in words of former US diplomat and Korea scholar Gregory Henderson, 'a state within a state, a vast shadowy world of … bureaucrats, intellectuals, agents and thugs.' By the 1970s the KCIA was targeting all levels of the population in all areas, since Park's control over society was almost absolute.[119] But in the 1970s, the KCIA's activities were often directed beyond the borders of its home country. Japan was an especially important target area, where there were numerous Koreans who spoke out against Park Chung Hee.

In congressional testimony, former KCIA chief Kim Hyung Wook testified about the Koreagate scandal, which involved South Korean attempts to illicitly influence members of the United States Congress, stating that Korean influence-buying networks in Japan existed on a far larger scale than those in the United States. Kim continued that there existed a secret agreement of bilateral cooperation in terms of exchanging information between the Japanese Police Agency and the KCIA. Although the Japanese Police Agency refuted this testimony, *Chongryun* responded by quickly denouncing the Japanese Police Agency, saying that the testimony of the former KCIA chief proved that Japan and South Korea tacitly cooperated with each other allowing Koreans who visited South Korea to study to be arrested as spies working for North Korea.[120] The accurate accounts still remain unavailable since the former KCIA Chief Kim Hyung Wook disappeared in Paris in October 1979, presumably murdered by the KCIA. But the number of people arrested who were visiting South Korea to study significantly increased, especially between 1975 and 1979.[121] The evidence, however, indicates that the target for restrictions over Korean residents became broader than before, no longer limited to the pro-North Korean diaspora group but extending to the subversive Korean residents with South Korean nationalities. The government began to keep an eye on Koreans' transnational activities, and the worry was that a second Korean War might be triggered. During the time, the everyday life of Zainichi Koreans was put under intense scrutiny by the Japanese Police Agency, whose main focus was on activities which threatened national security. Even casual meetings between pro-North and pro-South Koreans within the Zainichi were targeted by the Police Agency. This fact was highlighted by the one of the Zainichi Koreans I interviewed in Japan. Ms Park used to have her own business. After she had attended a business meeting with a Zainichi Korean who belonged to *Chongryun*, the police came to her offices and demanded to know the details. Although she was not involved in a serious case, she was not happy with such public intervention into her personal business transactions.

The democratization movement in South Korea intensified in the late 1970s, drawing much attention from young Korean residents, raising a diasporic consciousness among them. Facing a choice between the reality of their own position as permanent residents in Japan and diaspora Koreans backing democratization movement in South Korea, Japanese-born Koreans gradually adopted the expression *Zainichi* for their identification, giving themselves an attachment to the Korean homeland and distancing themselves from Japan.[122] Kang argues that the third or fourth generations' notion of being a permanently settled minority was a dangerous idea for the Zainichi itself. Kang claims that retaining links to the home country and their Korean identity would be the best way forward for the Zainichi.[123] By taking this path, the Zainichi would be able to highlight the injustice generated through discrimination. According to Kang, the Zainichi identity was a method of enhancing leverage in a state-centric world by abandoning discourse of temporary status but standing firm against integration into the host state. In the 1970s, especially from 1975 until 1979, a different category of Zainichi identity was created by the Japanese-born Korean diaspora. This identity was completely different from that of first-generation Koreans who supported either *Mindan* (pro-South Korea) or *Chongryun* (pro-North Korea) and embraced memories of their life in their homeland rather than in Japan. Eika Tai differentiates the identification between homeland-oriented Koreans – mainly those who support either *Mindan* or *Chongryun* – and Japan-born Koreans or homeland-oriented activists. Homeland-oriented Koreans paid a great deal of attention to the transnational nature of the colonial history while Japan-born Koreans and homeland-oriented activists were primarily concerned about oppression in the context of postcolonial Japan. The collective memory formed the emotional basis for the Zainichi identity, but the two groups of Zainichi Koreans carried that memory in different ways.[124]

Japan's nationality laws were a discretionary tool of the minister of justice, and could be withheld if officials judged the applicant to be undesirable to Japanese society, since only the acquisition of nationality guaranteed fully fledged citizenship rights in the 1970s, except for patrilineal intermarriage with Japanese.[125] Japanese nationality provided eligibility to apply for public-sector jobs, public housing and welfare benefits. The descent-based nationality laws of Japan excluded Koreans from achieving nationality automatically and kept them transnational, even though they had long since been assimilated in cultural, socio-economic or marital terms.[126] If Japanese policy makers viewed the transnational aspects of Zainichi Koreans as useful for enhancing Japan's own autonomy at the cost of security, they would provide accommodation temporarily while keeping an eye on transnational activity as they calculated the trade-off between security and autonomy under the asymmetric alliance. The 1975–9 period witnessed a time of strong cohesion between Japan and South Korea, which forced allied states to keep an eye on any subversive transnational activities on the part of the Zainichi diaspora that might enhance the North's revolution in the Korean peninsula whilst providing policies to regulate the ties between the Zainichi diaspora and its divided homeland.

Simultaneously, however, a new Zainichi identity emerged – one which maintained Korean nationality not merely as a *proof* of Korean ethnicity but as a *method* of enhancing leverage in a state-centric world by discarding discussions on temporary status and by sharing the collective memory of Japan's oppression in a post-colonial context. Decisions on naturalization did not only rest with the Japanese government but were also in the Zainichi's hands as well. By refusing naturalization, or by pursuing civil rights politics, the Zainichi maintained a permanent existence as resident aliens and retained an ambiguous identity until the optimal output was provided by Japan, the Koreas or other actors that had no nation state perspective. For the new Zainichi identity, Korean nationality may still provide an alternative essentialization of belonging to a homeland, but was no longer their sole means of expressing loyalty or aspiration. It was, however, a proof of the Zainichi identity. For those who had become culturally assimilated into Japan, Korean nationality became the last vestige of their ethnicity they could demonstrate,[127] and provided the Zainichi with an identity that has continually rejected rigid classification by challenging the very concept of the nation state.

Japanese public opinion and the country's media moved on to explore and expose the domestic conditions of the Zainichi in Japan. Many fully fledged non-citizen civil rights movements were fronted by Korean activists during the 1975–9 period. Although Korean identity and foreign legal status remained integral components of the movement, the emphasis was on the position of Koreans in Japan. As non-citizens without voting rights, Korean activists in collusion with Japanese activists demanded recognition and democratic inclusion based not on their status as nationals of another sovereign country but as rightful members of Japanese society.[128] As more Korean residents came to terms with the idea that they would live out their lives in Japan rather than return to the homeland,[129] the Japanese public and media started to pay more attention to the range of benefits that Japanese people enjoyed, such as welfare cover and national civil rights, and began to consider to what extent the Zainichi Koreans should be allowed to receive those same benefits and rights. During the 1975–9 period, Japanese public opinion and media influenced domestic areas of government policy towards the Zainichi rather than looking at transnational policies such as visa entry. However, the strict visa entry policy towards the pro-North Korean diaspora or Korean sympathizers to North Korean regime, as well as constant surveillance over their transnational activities in Japan, was driven by external geostrategic shifts connected to US foreign policy in East Asia rather than domestic considerations for Japan. With regard to the Japanese government's policy toward the Zainichi, alternative variables such as public opinion were not key factors of the policies that related to transnational space (such as visa entry and permanent status) which allowed the Zainichi to remain in limbo between host country and homeland.

Thus the 1975–9 period provided a crossroads at which the Zainichi bade a farewell to returning home and steered their way towards a future in Japan as a permanent foreign community. During the same period, US unilateral response following the fall of Vietnam allowed two satellite allies, Japan and South Korea, to come back into cooperative alliance, thus supporting the proposition that if alliance bloc cohesion is strong, the host state policy will be one of *exclusion* toward a diaspora group supported by an enemy state, but one of *inclusion* toward a diaspora group supported by an allied homeland (**Configuration 1**). The scope of an allied diaspora group, however, can be narrowed down by being stifled by the extreme surveillance being carried out in cooperation with a common allied homeland.

Weak Alliance Cohesion: Reagan's Strategy and a Return to Friction between Japan and the ROK (1980s)

Japan: Global player or Regional Partner?

President Ronald Reagan, after his inauguration on 20 January 1981, restored realist approaches to rationalizing the American moral crusade in the struggle between US perceptions of good and evil in international affairs. By rejecting Carter's liberal approach, Reagan asserted that the United States should worry less about its allies defecting and worry more about how it provoked misplaced opposition. By sharing the same beliefs as John F. Kennedy, Reagan claimed that 'if we cannot defend ourselves … then we cannot expect to prevail elsewhere … our credibility will collapse and our alliance will crumble.'[130] He was not interested in Nixon's strategy or his policy of détente towards the Soviet Union. The Reagan administration attempted to solve Western military deficiencies by spending more and more American dollars. Reagan's foreign policy objective was to reclaim strength and confidence through achieving across-the-board military superiority over rivals and to reassert US leadership through the reaffirmation and extension of security commitments to both traditional allies and anti-communist insurgencies.[131] He exerted relentless pressure on the Soviet Union as the 'Evil Empire,' increased the defence budget, started the Strategic Defence Initiative (SDI) and deployed intermediate-range missiles in Europe.[132]

The drastic changes brought about by Reagan's policies were also directed towards the other Asian ally, South Korea. George Shultz, who served as Reagan's secretary of state, recalled that to Ronald Reagan, South Korea was a stalwart ally and a vital symbol of resistance to communism.[133] Reagan did not view the US presence in Korea as an economic drain on the United States nor as an impediment to reducing tension on the peninsula as Carter had thought. Instead he saw the ROK as a loyal ally which had held the front-line in East Asia against the Soviet threat[134] and even performed the role of mentor to other less developed nations in the region.[135] Despite Chun Doo Hwan's military-dominated government, Reagan emphasized only the primary importance of the US–South Korean security relationship and played down issues of human rights and political liberation in South Korea.[136] If a major power provided reassurance and a large increase in security to its allied members, minor states would offer concessions to allow that major power to increase autonomy in exchange for security under an asymmetric alliance.[137] Deals between major and minor powers are natural in this situation. What was unusual during the 1980s was that friction did not only exist between major and minor powers but was also intense among minor powers, for example between Japan and South Korea.

As a clear signal of the firm alliance of the United States, the Reagan administration invited Chun as one of the first heads of state for summit talks with Reagan in February 1981. In his welcoming remarks to Chun at the White House, Reagan harkened back to the long history of close military cooperation between the two countries, praising General Douglas MacArthur's landing at Inchon and South Korea's participation in the Vietnam War and emphasizing 'our special bond of freedom and friendship' as well as mutual defence against aggression.[138] President Chun's domestic and international position was strengthened by the success of this visit, during which he also reaffirmed the critical importance of maintaining peace on the Korean Peninsula and in the Northeast Asia with President Reagan's cooperation and gained Reagan's support for South Korea's significant and on-going defence efforts.[139] By gaining US endorsement during Chung's visit, the implication was that South Korea's defence efforts were not only limited to defending itself but were also indirectly important to the security of Japan. Chung then started diplomatic negotiations to secure large-scale economic cooperation from Japan. Since South Korea had provided a buffer space to Japan for a long time, allowing Japan to escape direct threat from communist states, Japan had to offer some concessions to reduce the South Korea's heavy burden and take on a division of roles in the field of security. President Chung held an interview with Japanese *Jiji Press* on 15 May 1981, emphasizing his demand for Japan's economic cooperation based on an understanding that the two countries shared the same destinies. President Chung stated furthermore:

> An economic relationship between Korea and Japan is meaningful not only from the view point of economics but also from the implication that cooperation will bring peace to the Korean Peninsula together with peace of Northeast Asia which is directly connected to Japan, therefore, is expected to deliver

economic cooperation to Korea which will contribute to world peace and regional security.[140]

Foreign Minister Lho Shin-young officially presented a request for a $6 billion government Official Development Assistance (ODA) loan to finance South Korea's five-year economic development plan at the Foreign Minister's Conference between Japan and South Korea which was held in Tokyo in August 1981.[141] South Korea's economic growth in 1980 declined by 5.7 per cent, registering negative growth for the first time in decades. To finance the fifth Five-Year Plan, which envisaged a 70 per cent annual-growth rate, Korean planners calculated that at least $46.5 billion in foreign funds would be needed.[142] The Japanese were outraged at the South Korea's $6 billion proposal which amounted to more than half the total Japanese ODA budget for Asia and fourteen times its annual average outlay for Korea.[143] As far as the Japanese were concerned, the most unacceptable aspect of the demand was the link between security and the loan. Foreign Minister Sonoda rejected economic assistance from a security point of view, insisting that all the Japanese were willing to do was to make every effort to assist South Korea's civilian economy.[144] The links between security and the loan were clearly unacceptable to Japan, yet if Japan agreed to continue the negotiations in exchange for South Korea's relinquishment of the security links, the interests of both parties as well as those of the United States would be served. The deadlock, however, brought the negotiations a halt in the summer of 1982, when the textbook controversy arose, a controversy that lasted until Prime Minister Yasuhiro Nakasone came into office.

By early 1980, Korea was a society in turmoil, one in which the old constraints of the Park era were breaking down.[145] Friction between Japan and South Korea were also driven by generational changes; from an older generation of bureaucracy and a military which had been trained almost entirely by the Japanese during colonial era to new generation which could not speak Japanese and sought to establish a new tone of confidence and assertiveness over Tokyo.[146] Indeed, Chun and his entourage did not identify with the old elites' personal ties with Japan that grew from the colonial era,[147] but instead saw themselves as nationalistic leaders seeking to elevate Korea from its traditionally subordinate relationship with Japan.[148] However, like all dictators, Chun simply took over a vacuum Park had left and followed in his authoritarian footsteps. The Kwangju uprising of May 1980 began as a protest against Chun Doo Hwan's emerging military rule and grew into a full scale popular rebellion that included people from all strata of society.[149] For Chun, in order to strengthen his legitimacy as a ruler, he needed to present himself as a revolutionary, and anti-Japanese rhetoric became a convenient political instrument to enhance his political legitimacy in this regard. Textbook revisions did not only provoke outrage in Korea but also served as Chun's political tool to divert the Korean public's disaffection for his regime. In July 1982, Minister of Justice Seisuke Okuno stated publicly that avoidance of the term 'patriotism' in textbooks constituted a serious problem that needed correction. History writers were advised to moderate their account of Japan's responsibility for expansionist wars in the past and were told to substitute the term 'advance' for 'aggression.' According to the conservative leaders, it was improper for the young to be inculcated with feelings of guilt about their country.[150] In response to textbook revisions, Chun cancelled a scheduled Japan–ROK bi-national sporting event, postponed the autumn 1982 Japan–ROK parliamentary conference and threatened to recall his ambassador to Japan.[151] In the end, the Ministry of Education temporarily ended the textbook dispute by dropping its plans for revision.

The new Prime Minister, Ysuhiro Nakasone, set improved relations with the ROK as a top priority in 1983. Nakasone was known for his eloquent oratory and flair for showmanship. As the new premier of Japan, he personally telephoned his Korean counterpart, unlike past leaders, and proposed Seoul rather than Washington as his first overseas visit. The charismatic Nakasone delighted his hosts with his use of Korean phrases in speeches, his singing of Korean songs and his expressions of regret over the colonial past. The two leaders finalized the $4 billion loan agreement and declared a new era of bilateral cooperation.[152] Yet Nakasone's performance was actually well tailored towards his premiership ambitions as well. He also said that his top priority would be to improve relations with the US by resolving outstanding trade issues and increasing Japan's defence effort.[153] When Nakasone succeeded Suzuki as premier in 1982, Japanese diplomacy was in a state of confusion as far as the United States was concerned. In the midst of the increased level of international tension caused by the tough diplomacy of the Reagan administration, the Japanese government felt uneasy under the increasing pressure from the United States government that Japan should take greater and more clear-cut responsibility for East Asian affairs.[154]

Nakasone considered Japan–US relations the cornerstone of his security and his foreign policies, and saw the need to put a stop to what he considered a serious deterioration in that relationship over trade and defence issues. The Japanese Defence Agency's (JDA's) five-year defence plan (1983–7) increased the defence budget by 7 per cent and eventually raised total defence spending as a percentage of GNP to the 1 per cent ceiling. In 1983 Nakasone amended Japanese law to allow the transfer of Japanese military technology to the United States, and later joined in support for the SDI. To support US military contingency planning in the region in real terms, Japan expanded its capabilities to defend sea lanes of communication around the home islands. The Self-Defence Force (SDF) branches also took part for the first time in a series of joint military services with the United States to improve interoperability of forces.[155] Domestic opinion polls, however, indicated that public resistance to defence expenditures was increasing in Japan, with nearly two-thirds of the respondents favouring reductions. Public opposition and budgetary constraints were major factors in Nakasone's agreeing to only a 6.5 per cent increase for the 1983 fiscal year defence budget (excluding military pay raises), which would therefore not attain the level the US had been expecting.[156]

Concessions on major trade issues were another difficult task for Japan's dealings with the US. Anti-Japanese sentiment had become more widespread and pronounced in the United States at a time when Nakasone had taken over the Suzuki administration. In foreign trade, protectionist pressure in the US and Western Europe conflicted with domestic pressure for a government-aided export drive. Attempts to open the Japanese market further to imports were resisted by vested interests in Japan as well as by the bureaucrats who helped formulate the policies that had served so well in the past. Drawing the most widespread attention and media coverage in Japan was the issue of exchange rate stabilization. To the population at large – and especially the export-oriented businessmen hardest hit by the steep rise in the value of the yen – the attention given to the problem of stabilizing the dollar/yen exchange rate was indispensable for the Nakasone administration.[157] To make matters worse, strong resistance was driven by vested business interests and agriculture, both of which were important parts of the LDP's constituency. These domestic economic and political problems imposed strong constraints on Nakasone to fully respond to the demands of the US as well as demands from the minor power, South Korea, which under an asymmetric alliance formed a clear manifestation of the *weak* cohesion of the alliance.

However, the US Bureau of Intelligence and Research report worked out that Nakasone might be more independent and demanding than his predecessors on issues in which he differed from the US. Nakasone advocated an independent military capability for Japan, and criticized past policies that he felt had placed Japan in a subordinate role to the US.[158] In addition, the US already recognized that Nakasone had to maintain – and even enhance – his image as Japan's world statesman in order to strengthen his ability to function in office at a time of serious domestic political upheavals.[159] Nakasone had long believed that Japan should conduct itself with greater independence and self-confidence. Because of overwhelming Japanese public resistance to expanding military defence cooperation with US as a member of an alliance, yet still bearing aspirations for world leadership, Nakasone hoped to play the role of mediator in areas of international tension. For instance, he believed that he might be able to facilitate forward movement on the Korean Peninsula by brokering contacts between Seoul and Beijing and broadening unofficial Japanese channels to North Korea.[160]

The other difficult concession to reaching an agreement in dealing with a trade-off between Japan and South Korea was a policy consensus on North Korea. Based on the policy of promoting peaceful coexistence between South and North Korea, the Nakasone government was willing to gradually expand Japan's unofficial economic and cultural exchanges with North Korea in spite of South Korea's opposition to such a policy. Japan felt that the total isolation of North Korea from the international community could increase the possibility of irrational behaviour and actions on the part of Pyongyang.[161] The Nakasone government not only expressed its support for inter-Korean dialogue but also urged Korea's neighbours, China and the Soviet Union, to join Japan in cultivating an international environment conductive to the continuation of those talks. In addition, the Nakasone government advocated a modified version of the cross-recognition plan, which called for the recognition of Pyongyang and Seoul by Japan and China first, and then by the United States and the Soviet Union.[162] However, this was rejected not only by Pyongyang but also by Beijing and Moscow. Meanwhile, the Nakasone government's flexible policy toward North Korea aroused Seoul's strong protest. While the South Korean government no longer demanded recognition as Korea's sole legitimate government, it remained opposed to any official contact between Tokyo and Pyongyang, and remained fearful of any official contacts and trading connections that might strengthen North Korean military capabilities.[163] Despite Seoul's opposition, Japan decided to lift the sanctions it had imposed after the Rangoon incident with effect from 1 January 1985.[164]

Accommodation Policy to the Pro-North Korean Diaspora for Japan's Global Position

The numerous disagreements over trade, defence and diplomatic policy in terms of interests in an asymmetric alliance manifested a *weak* cohesion alliance, and under such a condition the host state is likely to *accommodate* the diaspora group supported by an enemy that is looking to expand its own autonomy (**Configuration 2**). In order to divert US pressures on trade and defence by highlighting actions in other areas, Japan tried to elevate its international status. North Korea was an area in which Nakasone hoped to play the role of mediator, easing the tension on the Korean peninsula.[165] The Korean diaspora card was again a useful instrument for broadening unofficial Japanese channels to North Korea. By reflecting on Japan's official policy that acknowledged the legitimacy of the division of the peninsula, the Japanese Ministry of Justice installed a new category of permanent residence in 1982, which they called Exceptional Permanent Residence, for Koreans (non-South Korea nationals) who had not qualified for permanent residence under the 1965 treaty.[166] Japan's international aspiration to world leadership status also contributed to a change in Japanese immigration policy. Responding to international policies on human rights, the Japanese government ratified the International Covenant on Human Rights in 1979 and the UN Refugee Convention in 1981. These ratifications forced the government to provide resident Koreans with access to the national pension programme and other welfare benefits, and to make it possible for all residents to apply for permanent residency status. The government expected that a more stable legal status and a policy of social integration would encourage resident Koreans to choose naturalization and allow them to be assimilated into the Japanese state.[167]

In fact, cultural socioeconomic assimilation accelerated after 1985, when government assimilation policy transformed the strict naturalization policy, extending patrilineal parentage to include matrilineal parentage. Coincidently, at around the same time, the number of much later Korean immigrants, who began to arrive in the mid-1980s in response to Japan's labour shortages, surpassed that of 'old-comer' Koreans – the Zainichi. The government started to see newcomers as more foreign than the Zainichi, regarding the old-comers as would-be Japanese nationals. As Brubaker argues, changing perspectives on immigration were not limited to Japan. Brubaker provides the evidence of a similar case of public policies in Germany as well. Like Japan, German citizenship law was well known for its restrictive attitudes towards non-German immigrants. What was less well-known is that except for political rights, long-settled non-citizen immigrants had possessed rights virtually identical to those of German citizens, which indicates that the solution of deep-rooted German 'differentialism' was made by extending the substantive rights of citizenship to immigrants without questioning their differentness, their foreignness, their 'otherness.'[168]

The new practices, policies and discourses surrounding citizenship, however, were assimilationist in the sense that they symbolically emphasize commonality rather than difference. Yet what Brubaker calls 'return assimilation' is not assimilation in the old sense, with its implications of complete absorption, but rather a more general and abstract assimilation, meaning to *become* similar. Rather than focusing on difference, the policy of return assimilation designates a direction of change rather than a degree of similarity.[169] To promote such a direction of change, or the gradual process of assimilation into Japanese society, the 1985 amendment to the Nationality Act finally eliminated the recommendation for naturalization applicants to adopt a Japanese name. The number of ethnically mixed Japanese nationals increased significantly in the 1980s as a reflection of these policies, which finally reaching a seventy per cent rate of intermarriage between Koreans residents and Japanese in the late 1980s.[170] (See Figure 4.2.) As Brubaker argues, Japan's new policies were assimilationist in the sense of politically recognizing, legally constituting and symbolically emphasizing commonality rather than difference, which

Naturalization statistics for Korean residents in Japan 1952–90

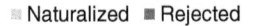

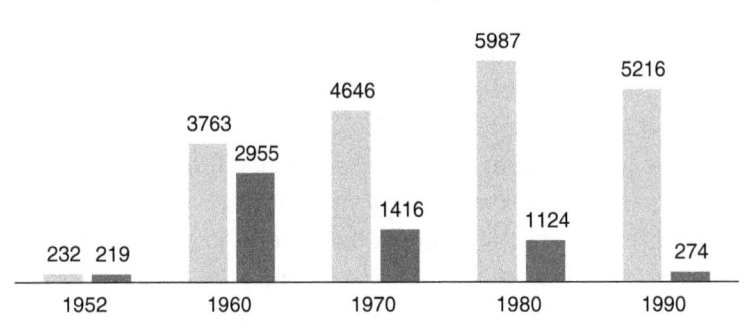

■ Naturalized ■ Rejected

Year	Naturalized	Rejected
1952	232	219
1960	3763	2955
1970	4646	1416
1980	5987	1124
1990	5216	274

Marriage ratio between Korean residents in Japan

— Marriage ratio

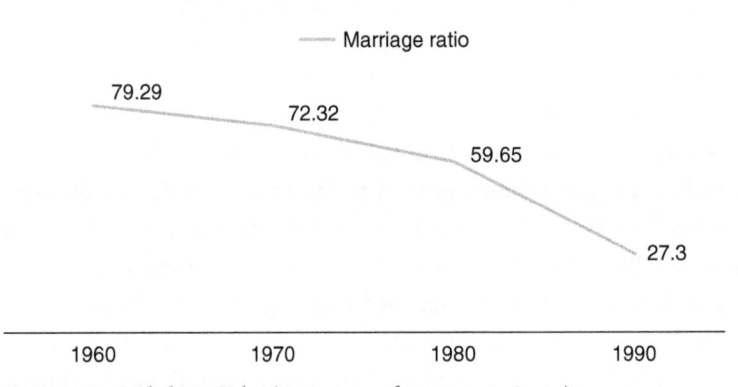

Year	Marriage ratio
1960	79.29
1970	72.32
1980	59.65
1990	27.3

Figure 4.2 *Gaikokujin Tokei* (statistics on foreigners in Japan)[a]
[a] Tai, 'Between Assimilation and Transnationalism,' p. 629: from Appendix: Statistics Koreans and foreign residents in Japan.

is how Mylonas defines an accommodation policy that retains the group in the state by maintaining the group's cultural specificities and institutionalizing its minority status. Thus it supports the proposition that if alliance bloc cohesion is weak, the host state (Japan) will accommodate a diaspora group supported by a homeland belonging to an enemy alliance bloc in order to expand its own autonomy over that diaspora group (**Configuration 2**).

The new category of permanent residence installed by the Japanese Ministry of Justice in 1982 allowed the pro-North Korean diaspora the right to receive single re-entry permits to Japan, which enabled them to travel abroad for the first time since the end of World War II, whether or not they possessed South Korean nationality.[171] More importantly, they could visit North Korea in order to be reunited with their repatriated families. Yet the passionate homeland politics of the first generation was no longer so vital to the pro-North Korean diaspora, which can be ascribed not only to generational changes within Japan's Korean communities but also to the improved standard of living for Koreans in Japan, which reflected Japan's long-lasting economic boom.[172] Indeed, once the pro-North Korean diaspora could actually visit North Korea, their interest in and sympathy for the so-called fatherland waned.[173] During the 1980s, the distance between the pro-North Korean diaspora and South Korea was reduced. In the past, telephoning relatives in South Korea had not been possible, as relatives in South Korea would be persecuted as the result of phone calls if those phone calls were intercepted or tapped by the South Korean authorities. In the late 1980s, however, such tensions seemed to vanish, allowing many first-generation Zainichi who identified themselves as pro-North Korean to recover their connections with family and relatives in South Korea.[174] In reality, it become more apparent to the pro-North Korean diaspora that North Korea would be a place to visit rather than a place to return to, as their lives were now rooted in Japan, where they worked and where their children were growing up. Such a feeling has been shared equally with subsequent generations from the other side of the Korean diaspora; it was impossible for the second or third generation to think of Kyongju, Seoul, Pusan or Cheju as their hometowns when they were actually born in Tokyo, Kyoto, Osaka, or Nagoya.[175] This was actualized as a result of a document issued by the Japanese Ministry of Justice in 1982.

In the view of Japanese elites, the pro-North Korean diaspora was no longer a target for deportation. Unlike previous Prime Ministers, Nakasone's capabilities as a statesman were not motivated only by economic or security concerns but also by Japan's potential elevation to world leadership. Indeed, during the economic surge of the 1980s it felt for a while as if Japan was going to become a world-leading economy. Japan's growing self-confidence as a world power embraced such aspirations as playing a key mediator role in the peaceful coexistence of the two Koreas, a role which might ease the frictions from either the US or South Korea. Prime

Minister Nakasone was very sensitive to any negative political fallout in the event that others – especially the United States or China – might negotiate an understanding without Japan's knowledge regarding movement in an inter-Korean thaw.[176] Nakasone wanted to take as much credit as possible for any progress achieved in interstate relations between South and North Korea in order to maintain Japan's good reputation as an intermediary.[177]

Interestingly, however, support for North-South dialogue did not preclude Japan from cultivating its own channels of contact with Pyongyang.[178] Japanese security was best served by an engaged, rather than an isolated, North Korea. Japan's increased military spending during the period, the expansion of its sea lane defence perimeter to one thousand miles and Nakasone's hawkish rhetoric served for his own revisionist foreign policies.[179] With no diplomatic relations between Japan and North Korea, however, the existence of *Chongryun* was the only available route between the two countries. For this purpose, the pro-North Korean diaspora was definitely an important card to hold. During the 1980s, Japan and South Korea still perceived different degrees of threat from North Korea.[180] Compared to South Korea, Japan's threat perception from North Korea was less intense, and their support for the diaspora group allowed the Japanese government to employ accommodation policies such as permanent residence grants to the pro-North Korean diaspora, despite the fact that it was supported by enemy homeland, in favour of its own revisionism foreign policy.

Japanese public opinion did not protest against government policy shifts towards the pro-North Korean diaspora, which came in conjunction with the widespread Korean boom in Japan, the surge in popularity of Korean music, language, and food that began in the mid-1980s.[181] Although South Korea showed little reciprocal interest in cultural exchanges, Japanese filmmakers became intrigued with the social plight of the Zainichi Korean community; formerly a taboo topic, and demand for books about Korea as well as about Korean perspectives on Japan grew.[182] At the same time, Tokyo undertook an expansion of contact with Pyongyang, particularly in the political arena, and a number of prominent Japanese politicians discussed issues such as the establishment of trade liaison offices and the expansion of political exchanges (in the past any exchanges had been mostly confined to the educational and cultural spheres).[183] For this, *Chongryun*'s presence provided a link with the North Korean government.

Conclusion

This chapter explained that Mylonas' proposition is applicable only under conditions of strong cohesion within asymmetric alliance bloc. If alliance cohesion is strong under asymmetric relations, a host state is likely to accommodate a diaspora group supported by an allied state. However, this chapter also favours the idea of a causal nexus between alliance cohesion and diaspora for nation-building policies, which explains that if the alliance cohesion is weak, a host state accommodates the diaspora group supported by an enemy due to its desire to expand its own autonomy under an asymmetric alliance. If we look at domestic factors such as media or public opinion, it has not greatly affected the area of Japanese government policies such as visa entry or permanent status regarding the transnational space in which the Zainichi existed during the Cold War. In other words, these policies have been effectively controlled by Japanese elites who consider Japan's own geopolitical strategy, which is driven by external geostrategic shifts connected to US foreign policy in East Asia rather than domestic considerations for Japan.

Another important point this chapter has identified is the fact that the degree of threat perception by the host state also should be taken into account. During the Cold War, the interpretation of the 'Korea clause' was a vital measurement of allied cohesion. As long as the Korea clause continued to emphasize the security of South Korea as being essential to Japan, two governments constituted the closest approximation to a defence treaty between Japan and South Korea.

But why did Japan fail to acknowledge the severity of the North Korean threat to South Korea? One reason for this is the role of the United States, which allowed the Japanese to minimize that threat perception. Japan returned to a policy of strengthening ties with South Korea only when the US commitment to the Asia–Pacific region became weak (exemplified by Carter's plan to withdraw US troops from the Korean peninsula). During the Cold War era, East Asian regional geopolitics had been so much dominated by US ideology, which sometimes permitted Japan to ignore the geopolitical consequences – something that Carol Gluck referred to as 'geopolitical blindness.'[184] Without a US presence, the control of the Yellow Sea, the Sea of Japan and exit routes for Soviet naval bases would become strategic concerns suddenly thrust upon Japan.[185] Based on this logic, it might be possible that the threat perception by the host state is not necessarily correlated with alliance cohesion outside the realm of US-centric geopolitics. After the end of the Cold War, the Japan–US relationship was no longer the only relationship that mattered. Japan had increasingly to deal with Asia and China and their demands. The post-Cold War years had to face the memories of war by the early 1990s due to geopolitical shifts. This was not an easy task for Japan, since the role of the United States had frozen Japanese memories of the war for a long time. The problem was that the memory would reassign a stigmatized identity to the Zainichi diaspora.

5

Does Alliance Cohesion Still Matter in the New Post-Cold War (1990–2014)?

Introduction

During the Cold War, the Japan–US alliance contributed to fostering friendly and cooperative relations between Japan and South Korea, despite antagonism over historical issues. Although common security and economic interests were basic factors that bound Japan and South Korea, the extent and strength of these ties cannot be defined with any logical coherence. Alliance cohesion is tested when critical exogenous geopolitical factors change, and the space occupied by the diaspora moves in consequence, either by expanding or contracting in relation to the policies employed by the host state: exclusion, assimilation or accommodation. With the end of the Cold War, however, new geopolitical power configurations have made the strategic balance in North Asia more complex. As Brett Ashley Leeds and Burcu Savun have found, alliances are more likely to be abrogated on an ad hoc basis when one or more members experience changes that affect the value of the alliance, such as a change in international power, a change in domestic political institutions or the formation of a new external alliance.[1]

The rise of China as an economic superpower and the North Korean threat are indispensable factors which have inevitably reconstituted the value of the alliance and present challenges for Japan and South Korea in terms of security and cooperation between Japan and South Korea in the context of the US–Japan and US–ROK security alliances. First of all, the question of China is likely to affect not just the economic but also the political influences that are at work on the Korean peninsula. During the Cold War the South Korean government had a quasi-alliance with Japan by virtue of the fact that both South Korea and Japan were allies of the United States.[2] With the collapse of the Soviet Union, however, and the establishment of deep economic and political ties between South Korea and China following the normalization of diplomatic relations between the two countries in 1992, the threat China presents can no longer be taken as a given, although the differences in the South Korean and Japanese responses to China's growth have been more striking than the similarities.[3] South Korea has nurtured a new relationship with China and now plays a more influential role as a middle power in East Asia, as manifested in the ASEAN Plus-Three and the East Asia community processes.[4]

From the Japanese point of view, however, the possibility of a Korean peninsula dominated by Chinese influence is no less welcome to Japanese security partners. In this new post-Cold War geopolitical environment, a weakening of the US defence commitment does not necessarily lead to greater cooperation between Japan and South Korea, which shows that Cha's theory of Quasi-Alliance does not work in the post-Cold War era. China has taken over from the US as the both Japan and South Korea's biggest trading partner, which implies a major change in power relations.[5] Thus the shifting balance of power on the East Asian sub-continent and the creation of a positive overall relationship between South Korea and China will present a bigger challenge for Japan and the Japan–US alliance in the years to come.[6] Recently, some scholars have argued that the formation of a new alliance between China and South Korea is only a matter of time.[7] Indeed, since the end of Cold War, Chinese influences have begun to affect not only the economic balance but also the security balance, as South Korea gradually sought its own role between the rising power of China and the traditional alliance with the United States.[8]

Additionally, from the Japanese standpoint, the North Korean threat has become a more prominent concern for Japan in the post-Cold War years than it was in the Cold War era. In particular, North Korea's development of nuclear and ballistic missile programmes have increased the North Korean threat from a local one on the Korean peninsula to a regional and global one by the development and proliferation of weapons of mass destruction.[9] In the Cold War era, North Korean missiles could not reach Japan, but by the 1990s – particularly in the wake of the 1993 Nodong and 1998 Taepodong missile tests – the North Korean threat had become an real and immediate threat to the Japanese homeland.

During the Cold War era, North Korean factors acted as bargaining chips between South Korea and Japan. Thus for Japan, the pro-North Korean diaspora was a useful political instrument, demonstrating one of this book's most important arguments – that weak alliance cohesion allows the host country to accommodate a diaspora supported by enemy allied homeland in order to expand its own autonomy under an asymmetric alliance. Post-Cold War geopolitics, however, have brought the host state and the divided homeland into new geopolitical configurations, which inevitably affect the space occupied by the Zainichi in terms of the policies of exclusion, assimilation or accommodation employed by the host state. In the post-Cold War years, as the North Korean threat becomes a more independent factor, the degree of threat perception of the host state has a strong impact on policy making toward the diaspora.

As the Cold War went on, Japan failed to acknowledge the severity of the North Korean threat to South Korea. One reason for this was the role of the United States, which allowed the Japanese to minimize its own threat perception. Japan turned its policies towards strengthening ties with South Korea only when the US commitment to the Asia–Pacific region became weak. In the post-Cold War era, however, as American experts have noted, North Korean missile tests were a possible indicator that any subsequent Korean War might not occur on the Korean peninsula, but instead on Japanese soil.[10] After all, East Asian regional geopolitics in the Cold War era had been strongly dominated by US geopolitical strategies, and this sometimes allowed Japan to ignore geopolitical consequences and conduct bargaining deals between autonomy and security under the strong US security umbrella. As a result, the space occupied by the diaspora shifts either by expanding or contracting in reflection to the policies employed by the host state; exclusion, assimilation or accommodation. In the post-Cold War years, however, the basic consensus that pertained during the Cold War has been challenged; strategic changes such as the rise of China, the comparative decline of Japan and the weakening of American hegemony have significantly affected the Japan–US alliance relations on the Korean peninsula.[11] Based on this logic, it might be possible that the threat perception by the host state does not necessarily correlate with alliance cohesion in the absence of US-centric geopolitics. So the problem facing the diaspora now becomes a question of the conditions under which their host state will treat them – inclusively or exclusively – in refection to the degree of alliance cohesion in the post-Cold War years.

Brett Ashley Leeds and Burcu Savun identify four factors crucial to the value of alliances: the level of external threat faced by the allies, the military capabilities of the allied states, the extent to which policy goals are shared by the allies and the availability of substitute allies. If we trace the alliance cohesion during the Post-Cold War years, despite the warning that the Japanese–US alliance was going adrift after the end of the Cold War, North Korean threat factors have become a catalyst for the consolidation of the Japan–US alliance and the deepening of security ties with South Korea from the 1990s until 2002. However, from 2002 to the present, the policy responses to the North Korean threat have differed between the United States, Japan and South Korea, and the rising China has gradually become an obstacle to alliance consolidation. By focusing on the four factors crucial to the values of alliances Leeds and Savun provide, I divide the Post-Cold War years into years of weak Alliance cohesion (1990–2), then the stronger years (1993–2002) followed by a return to weak Alliance cohesion from the autumn of 2002 until now. In the light of those divisions, I can then test a set of hypotheses that would predict possible policies towards the diaspora in relation to alliance cohesion between host country and homeland. When conditions change, as they did with the end of the Cold War, the host state treats diaspora groups in a different way to that suggested by the hypothesis, and will now depend on additional factors related to the limitations of correlation between alliance cohesion and diaspora as a category. More specifically, I pay attention to security concerns such as the North Korean threat and the rise of China. The chapter concludes by examining the conditions under which the host state treats the diaspora as a reflection of shifts in alliance cohesion.

The Post-Cold War Era (1990–2014)

The Shift from Weak Alliance to Strong Alliance Cohesion: Alliance Cohesion Still Matters (1990–2002)

With the Cold War over, Japan faced a challenge to extend itself beyond the old framework of the asymmetric alliance. The essential question for Japanese policy was raised by the collapse of the Soviet Union – what could be made of Japan's global role in the years after the Cold War? As the Cold War ended the US Congress began questioning the value of its asymmetric alliance with Japan. When the 1990 Gulf Crisis occurred, the value of the asymmetric US–Japan alliance was tested; would Japan prove to be a reliable ally to the United States? A wave of fear of abandonment in the Post-Cold War era swept over Japan.[12] Tokyo had been severely criticized in Washington for the lack of personnel contribution to the Gulf War despite Japan's financial contribution of 13 billion dollars. US Congress strongly questioned the value of the US–Japan alliance, asking whether Japan could offer any more support for the provision of bases for US forces in case of contingency on the Korean Peninsula.[13] The Korean peninsula has always occupied a high priority as an issue in the US–Japan alliance. Since the end of Cold War, however, the relationship between Korea and the US–Japan alliance has gradually shifted over time as changes occurred in the circumstances on the peninsula.[14] Although Morrow (1991) claims that asymmetric alliances last longer than symmetric ones, his argument assumes that alliances between major and minor powers are based on implicit issue linkage – that major powers provide security to minor powers in return for support on other issues.[15] By investigating the four factors related to alliance values provided by Leeds and Savun – threat level, military capability, shared policy goals and the availability of substitute allies – this chapter assesses the degree of alliance cohesion, then identifies how this cohesion affects policy making towards the two Korean diaspora groups; one supported by a common allied homeland and the other by an enemy allied homeland.

Assessment of the Value of the Alliance: Military Capability
Alliance is useful for both deterrence and compliance. Credible alliances allow states to accomplish their national security and foreign policy goals

more effectively and efficiently. In order to do this, leaders are inevitably expected to cover potential costs that full alliance obligations incur; this might involve a choice between participating in a war or abandoning an ally in its hour of need.[16] Such credibility relies on military capability, and Japan had several impediments as a fully reliable security partner for either Korea or the United States under this heading. According to Hisahiko Okazaki, a former Japanese Foreign Ministry diplomat, the main impediment was Japan's self-imposed domestic considerations under its own Constitution, which made Japan a less than reliable security partner for either Korea or the United States.[17] Alliance with the United States highlighted a dilemma that minor states aligned with major states have experienced since Thucydides and the Peloponnesian Wars – the dilemma between entrapment and abandonment. In order to escape this dilemma and empower themselves within an alliance, minor states can establish their own military capabilities, broaden relations with other countries or strengthen their own economic capabilities.[18] None of these solutions was applicable to Japan's dilemma in the earlier post-Cold War years, in particular 1990–2002.

Japan's policy of restraint in the use of military force is based on the present interpretation of Article IX of the Japanese Constitution, and its non-nuclear principles also make the alliance with the United States, including the nuclear umbrella, critical.[19] Additionally, neighbouring countries' concerns, especially South Korea and China, over Japan's military role in regional and global security persist due to the nightmare of World War II and memories of Japan's militarism in the 1930s and 1940s.[20] Such concerns also limit the deepening of relationships with other countries in Asia. The fact that the revision of the Japanese Constitution is viewed in China and Korea as a dangerous taboo suggests that any expansion of Japanese cooperation with neighbouring countries in Asia still comes up against an emotional barrier due to deep-rooted historical antipathies. Finally, the collapse of Japan's bubble economy wreaked considerable havoc on the Japanese worldview. When the economic boom collapsed in 1990, many argued that the problem was temporary and not structural, but by the end of 1997 Japan had seen five consecutive quarters of negative growth. Then for the first two quarters of 1999, Japanese GDP grew at 1.5 per cent, but as public spending flagged in the third quarter of 1999 the economy continued to contract.[21] During the 1990s, Washington gradually became concerned by the growth of Beijing's military power, although it had not reached the level of a potential long-term competitor on the level of the United States. President Clinton's heralding of a new strategic partnership with China in 1998 confirmed to a number of Japanese observers the trend of 'Japan passing.'[22]

In the early post-Cold War years between 1990 and 2002, the assessment of military capability as an alliance value put the Japanese stance into a state of unresolved dilemma between autonomy and security, since a minor ally does not resolve the fundamental military dependence on the major ally unless the major ally becomes economically dependent on the minor ally without finding or developing relations with other countries. The only way for Japan to escape the situation was to increase its own autonomy by changing its Constitution, which meant moving away from its existing allegiances. In the early post-Cold War years, however, it was still difficult to support constitutional revision from either Japan's internal or international – especially Asian – regional consensus.

Assessment of the Value of the Alliance: Shared Goals
Having said all this, leaders are likely to value allies with similar policy goals to their own. Gaubatz Kurt Taylor argues that alliances between liberal democratic states have proved more durable than either alliances between non-democratic states or alliances between democratic and non-democratic states.[23] Because democracies are societies based heavily on respect for the rule of law, they are unlikely to abrogate existing commitments which have become institutionalized as part of domestic or international legislation.[24] Given the changes in the balance of power in the East Asian region – the reduction of US hegemony, the relative decline of Japan's economic power and the rise of China – the United States sought broader partnerships in the provision of public goods with like-minded nations including Japan and South Korea. Although Japan and South Korea have to deal with their own political differences over history, US strategy increasingly framed the alliances between each country in similar terms and in a global context by providing the foundations for enhanced trilateral policy coordination.[25] Michael Auslin and Christopher Griffin have argued that trilateral coordination between the US, Japan and South Korea

might be bolstered by the establishment of a Trilateral Security Committee which would affirm and guide working-level negotiations among the three countries based on a common strategic vision and focused on cooperation for humanitarian disasters, cooperative maritime security and missile defence.[26] Furthermore, in the context of such a trilateral nexus, Ralph Cossa advocates that the US, Japan and Korea should work toward forming a 'virtual alliance' which would contribute to developing a close, cooperative, strategic triangular relationship between them.[27]

However, this is not easy task because the challenges to effective political management of history and territory issues between Japan and South Korea have been a continuous source of friction in Japan–ROK relations, preventing deeper levels of cooperation as a practical possibility after the end of the Cold War. At this point, the Japan–US relationship was no longer the only relationship that mattered. Japan was increasingly forced to deal with Asia and China and their demands, which compelled the Japanese government to face memories of war and unresolved territorial issues. In 1996, Japanese and South Korean preparations to ratify the United Nations Convention on the Law of the Sea (UNCLOS) led to competing claims to a small group of islands situated between the two countries. The Takeshima/Tokdo Islands became crucial symbols of each nation's exclusive economic zone (EEZ) under the UNCLOS rules. In February 1996 the South Korean government began building a wharf on one of the islands and Korean President Kim Young Sam appeared on national television to demonstrate his commitment to defending the islands, an event that further inflamed Korean and Japanese public sentiments.[28]

In terms of historical issues, questions about comfort women[29] began to surface in the Japanese Diet in 1990, and in 1993 Chief Cabinet Secretary Yohei Kono issued a report acknowledging the responsibility of the Imperial Army, offering his heartfelt apologies and expressing deep remorse to all those affected, a statement which received a number of objections by Nationalists in the LDP.[30] The Kim Young Sam government immediately protested Japan's inadequate handling of the issue and demanded compensation. The Japanese government did not offer compensation directly to the victims, but instead established a private Asian Women's Fund, which distributed $17,000 to each victim with a private letter from Prime Minister Hashimoto.[31] Japan had struggled with the apology issue since the end of the Cold War in the 1990s without any success.[32] Carol Gluck argues that when Koreans assert that Takeshima is not Japanese territory, it is now common for them to bring up the comfort women issue, which has become part of the geopolitics of memory. The seismic geopolitical shift that followed the end of the Cold War heralded change but at the same time brought uncertainty and drove nations both old and new to review their recent history in a new and more nationalistic way. Gluck also adds that an uncertain, changing world order, mixed with economic difficulties, is a recipe for rising nationalism.[33]

Nonetheless, a newer liberal democratic concept – one that had been eschewed by newly elected President and symbolic leader of Korea's democratization movement Kim Dae Jung – allowed the South Korean government to pave the way for a new escape from the historical and territorial problems that had dogged the relationship between the two governments. During Kim Dae Jung's state visit to Tokyo in October 1998, he and Japanese Prime Minister Obuchi went to work on a draft joint declaration that would feature a formal Japanese apology based on a formula Kim Dae Jung proposed; Japan would apologize and Korea would accept that apology once and for all while expressing appreciation for Japan's international role.[34] Following this agreement, Kim took the steps to ease the way for an apology from Japan by unilaterally compensating comfort women, referring officially to Akihito as the emperor for the first time in ROK history and agreeing to lift the long-standing ban on Japanese cultural imports into Korea.[35]

Japanese financial assistance during South Korea's economic crisis in late 1997 also contributed to promoting a sense of mutual dependence and cooperation rather than acrimony and conflict between Japan and Korea. Under the 1998 Kim-Obuchi joint declaration, the Japan Bank for International Cooperation signed a memorandum of understanding with Korea's Ministry of Finance to provide an additional $3 billion in united loans for small to medium-size South Korean firms.[36] Such mutual cooperation reflected cooperation in security issues that had deepened among the allies throughout the 1990s. The 1998 Japan–ROK Joint Declaration also announced a New Japan–ROK partnership for the twenty-first century; alliance with the United States was affirmed and the promotion of security dialogues and defence exchanges was supported. Security dialogue, including defence exchanges at various levels between Japan and the ROK, had been actively promoted by the Japanese Defence Agency since the mid-1990s, and were eventually to lead to the first joint Japan–ROK naval exercise, held in August 1999.[37] The peace and stability of the Asia–Pacific region are common shared goals among allied members who have similar values and support the ideals of liberal democracy. During the earlier stages of the post-Cold War era (1990–2002), those shared goals contributed to strengthening and consolidating the cohesion of the alliances between the US, Japan and South Korea. In particular, the years from 1998 to 2002 represented a time of thawing relationships and cooperation for Japan and Korea, from the Joint Declaration to the co-hosting of the 2002 World Cup.

Assessment of the Value of the Alliance: Threat Level

Leaders are far more likely to value alliances when they face external threats.[38] This is one of the key factors that make alliances valuable. As Morrow argues, alliances signal to external parties the willingness of allies to come to one another's aid if threatened by other nations. Such signals could enhance threat deterrence by convincing threatening nations that intervention against them was likely from an alliance of countries, not just the threatened nation itself.[39] Such a show of strength allows alliances to coordinate their war plans and foreign policies as well. The coordination of foreign policies means that one or all of the allies must abandon their own nationalist policies that seek individual countries' desired ends and instead prioritize the elimination of conflicts of interest that could separate the allies. External threat is therefore an indispensable factor in consolidating alliances. Friction between Japan and South Korea emerged in the immediate years after the end of the Cold War, yet the rise of the North Korean nuclear issue forced Japan and the ROK to work together and cooperate with the United States. Although experts warned that the Japan–US alliance was going adrift in the years immediately after the end of the Cold War, the North Korea nuclear crisis of 1993–4 shifted attention back to Korea and regional security.[40] As a result, the alliance between the US and Japan became further consolidated by the end of the 1990s in terms of defence cooperation, non-proliferation diplomacy and trilateral cooperation with South Korea.

During the Cold War, Japanese politicians frequently attempted to manoeuvre deftly between the two Koreas without abandoning Japan's *de facto* alignment with the South. The common cause of such manoeuvres between Japan and South Korea was Japan's shifting diplomacy policies towards the North under the weak alliance cohesion, as I explained in the previous chapter. This factor in Japan–ROK relations became particularly complicated after the end of the Cold War, with Japan's new diplomacy toward the North. In 1990, influential LDP politician Kanemaru Shin met with Kim Il Sung and vowed in a tearful statement to normalize relations with the North as soon as possible. Kanemaru's promise of huge reparations to the North, however, had threatened to undermine Seoul's own negotiating strategy. The South Korean government warned Japan to slow down, since new hope for reenergizing the normalization process emerged in December 1991, when South and North Korea completed the Agreement on Reconciliation.[41]

Changes in domestic Japanese politics, however, undermined the prospects for normalization with North Korea. In 1992, Japanese media discovered North Korean provocations during a confession from a North Korean commando, Kim Hyon Hui, who stated that North Korea kidnapped innocent Japanese citizens to teach spies in the North. By 1992, Japanese police investigations of missing persons along the Sea of Japan coastline pointed to North Korea's intelligence service as the likely abductors of many people. Pyongyang denied this allegation and refused to cooperate on the issue. In response to the issue of emerging Japanese abductees, the Japanese public reacted with shock and gradually began to perceive North Korea as a growing threat. The Japanese Foreign Ministry, facing intense domestic pressure, walked out of talks with North Korea on 4 November 1992. In November 1992, Kanemaru fell from power and was arrested for tax evasion in the Sagawa Kyubin scandal.[42]

Furthermore, as the Kim Il Sung regime grew more and more isolated from its sponsors in Moscow and Beijing (although Seoul successfully pursued normalization ties between the countries), North Korea increasingly resorted to infiltration, limited armed conflict and development of weapons of mass destruction (WMD) for its own survival.[43] In March 1993 North Korea abandoned the Non-Proliferation Treaty (NPT). Then in May 1993, the North launched a Nodong-1 SCUD-type missile, which had a strike range that included Japan. Indeed, it was the North Korean threat that became the catalyst for enhanced security cooperation between the allies in the 1990s. As Morrow explains, there was a strategic convergence which required the coordination of foreign policies toward the North Korean nuclear problem between the Japan–US and ROK–US Northeast Asian alliances and priorities were aligned to focus on the nuclear issue.

Harmonious coordination among allied members was tested when the US–DPRK Agreed Framework and the Korea Energy Development Organization (KEDO) were established in October 1994 and March 1995, respectively. These became the foundations for bilateral and trilateral cooperation with the ROK on the denuclearization of North Korea throughout the 1990s. The Agreed Framework was a non-proliferation initiative in which the Clinton administration agreed to provide energy assistance and improve US–DPRK bilateral relations in return for North Korea's adherence to the Nuclear NPT. In order to provide energy assistance, Washington asked Seoul and Tokyo to share the burden. Washington would provide heavy fuel oil (500,000 tons annually) and Seoul and Tokyo would provide funds for a Light Water Reactor (LWR) project to be completed by 2003.[44] In the earlier process of establishing the Agreed Framework, Japan and ROK were very reluctant because the framework did not address North Korea's pre-1994 military capabilities, nor did it address the issue of ballistic missiles, which was a primary concern for the Japanese. In the end, however, both Japan and South Korea decided to cooperate with the Clinton administration by focusing on capping the North Korean nuclear programme for now and the future.[45]

However, Japan found itself ousted from diplomacy on the peninsula. When President Clinton and South Korean President Kim Young Sam met on Cheju Island in April 1996, they proposed Four Party talks to bring together China, the DPRK, the United States and the ROK to work toward a permanent peace treaty on the peninsula to replace the armistice of 1952.[46] Neither Japan nor Russia was included in the negotiations. However, Japan saw room for another attempt at normalization talks with Pyongyang in 1996. The parallel development between the Four Party talks and Japan–DPRK talks continued, but since opposition to normalization with North Korea hardened within the LDP after struggling to get domestic consensus, Japan's direct negotiating pipeline to the North was disrupted by the

North's development of more powerful ballistic missiles, rather than strong domestic opposition. On 31 August 1997, North Korea launched a three-stage Taepo-dong ballistic missile directly over Japanese airspace. After the Taepo-dong launch, the Japanese government found new enthusiasm for security cooperation with Seoul, and a trilateral defence dialogue moved from ad hoc unofficial sessions in the early 1990s to regular official meetings. The Ministries of Defence of Japan and South Korea agreed to hold joint naval search-and-rescue (SAR) exercises and establish a bilateral hotline for crisis management.[47] More importantly, in October 1998 Japan and South Korea announced a Japan–ROK joint declaration which pledged to increase defence exchanges and consultations and establish regular bilateral cabinet meetings.[48]

Summary of the Assessments on the Alliance Cohesion (1990–2002)

Thus, the post-Cold War years of the 1990s and early 2000s saw a shifting pathway – from going adrift in the immediate years that followed the end of Cold War (1990–2) to consolidation of the Japan–US alliance and the deepening of security ties with South Korea (1993–2002), which emerged primarily as a result of the North Korean nuclear issue (1993–4). Although historical and territorial issues flared in February 1996 during the Kim Young Sam administration, and could have undermined security cooperation, the two countries managed to continue their dialogue on security issues. Also, during the negotiation process for the Agreed Framework, Kim Young-sam appeared to seek the availability of preferred substitutes by proposing Four Party talks to bring in China instead of Japan. In the early stage of the post-Cold War era, however, the rising power of China had not reached the level at which it might affect global leaders' views on the value of existing alliances. In order to coordinate defence policy, liberal democratic values have allowed alliances with higher levels of military institutionalization, evidenced in the wake of Kim Dae Jung's election as ROK President. Alliances could now operate as signals of common interest for the allies. The coordination of foreign policies was required for each alliance member to achieve that common interest, which would eliminate pre-existing conflicts of interest that might otherwise have separated the allies. In the early post-Cold War years, from the 1990s until 2002, the common interest was peace and stability within the north-east Asian region. In order to achieve this shared goal, the North Korean nuclear threat resulted in tighter allegiances that worked more effectively together, while the coordination imposed some cost on allies by compelling them to lower the priority of individual interests regarding their foreign policies that might separate the allies.

Alliance Cohesion and the Diaspora in the Post-Cold War Era (1990–2002)

It was a coincidence that the year the Cold War ended corresponded closely with 'the 1991 question' over the continuing legal status of future-generation Koreans born to Korean parents holding treaty-based permanent residence status. This status was due to expire in 1991,[49] and as the date of expiration approached, a growing anxiety was expressed by the Zainichi that unless the treaty-based rights were extended to future generations, their children would lose the right of permanent residency. The South Korean government had been urging the Japanese government to re-negotiate this issue since 1985.[50] The Japanese Ministry of Justice had installed a new category of permanent residence in 1982 called Exceptional Permanent Residence, which was offered to Koreans who had not been eligible to obtain permanent residence under the 1965 treaty – that is, the pro-Pyongyang diaspora. So, as 1991 approached, the Ministry began issuing single re-entry permits to Japan, which enabled *Chongryun* Koreans to travel abroad after three and a half decades of confinement within the archipelago.[51] In 1990 there were about 268,000 holders of Exceptional Permanent Residence status, while the number of holders of permanent residence under the 1965 treaty amounted to about 323,000. Indeed, 'the 1991 question' was influenced by the fall of Eastern European regimes, the end of the Cold War, and the shift in the global balance of power, all of which occurred at roughly the same time.

In the years immediately after the end of the Cold War (1990–2) and before the North Korean nuclear crisis of 1993–4, the Japan–US alliance was going adrift due to shifts in the global balance of power. Without the Soviet nuclear threat, security affairs in East Asia were no longer perceived as directly connected to the safety of the United States, and the bases provided by Japan were not considered as significant as they had once been.[52]

Meanwhile, Seoul's *nordpolitik* and the pursuit of ties with Moscow and Beijing also went ahead during the same period.[53] Given such a new geopolitical space after the immediate end of the Cold War, it might be reasonable to assume that Japan was hoping to capitalize on thawing conditions in Northeast Asia to improve ties with the North. During this period, a shift in the global balance of power also created a new regional geopolitical space, which inevitably weakened the alliance cohesion on a temporary basis. Thus the period between 1990 and 1992 was an era of transition categorized as a time of weak alliance cohesion, and thus applicable for **Configuration 2**: if alliance bloc cohesion is weak, the host state (Japan) will accommodate a diaspora group supported by a homeland belonging to an enemy alliance bloc in order to expand its own autonomy.

Meanwhile, during the Cold War, Japan maintained informal ties with the North by using the diaspora, the pro-Pyongyang General Association of Korean Residents in Japan (*Chongryun*), as a bargaining tool. From 1990 to 1992, the Japanese government also granted the status of Special Permanent Residence to those Koreans and their offspring who had lost their Japanese nationality in accordance with the Peace Treaty, in order to use the diaspora card as a political tool to stretch across the space between state boundaries to encompass future capabilities. When Prime Minister Kaifu visited the Republic of Korea in January 1991, a memorandum was signed by the Foreign Ministers of the two countries.[54] Then in 1992, all Korean permanent residents – including most *Chongryun* Koreans – were made special permanent residents regardless of whether their alien registration card bore *Chosen* or *Kankoku* (Republic of Korea). They became eligible to apply for various social benefits including re-entry permits to Japan.[55] So in order to solve the 1991 question, the Japanese government, on 10 May 1991, enacted 'the Special Law on the Immigration Control of Those Who Have Lost Japanese Nationality and Others on the Basis of the Peace Treaty with Japan,' by which all third- and fourth-generation Zainichi Koreans were to be guaranteed the right of permanent residency in Japan, regardless of their pro-South or pro-North affiliations. Indeed, the weakness of Cold War strategies at that point allowed the Japanese government to apply this new law to all Korean residents, including members of the diaspora supported by North Korea. This can also be held up as proof of **Configuration 2** of the alliance theory.

The socio-economic position of the Zainichi Koreans significantly improved in the 1980s and 1990s. Due to the housing bubble economy during these years, the expansion of the service and construction industries provided Koreans – who were concentrated in these industries – with various job and business opportunities. One of the typical successful businesses was the *pachinko* industry[56], considered 'the Koreans' biggest ethnic industry in Japan.'[57] In 1994, according to the Administrative Management Bureau, the pachinko industry was worth more than 30 trillion yen per year, a figure that was twice the value of Japan's domestic automobile industry at that time.[58] Remarkably, the pachinko industry was largely rooted in the pro-North Korean diaspora community, which was sending an estimated $600 million per year back home to the North.[59] In addition, there were considerable political funds flowing to the politicians of both the ruling party, the LDP, and of the opposition party, the JSP (Social Democratic Party of Japan) from the pachinko industry as well.[60] As a matter of fact, in the early years of the post-Cold War era (1990–2), in parallel to the diasporas' private remittances, Japan and North Korea engaged in about $500 million per year in trade – an amount that was growing in relative importance to the North, since funds from the Communist bloc had already dried up.[61] Indeed, the policy of accommodation for both pro-South and pro-North Korean diaspora groups was reasonably employed in response to the new post-Cold War geopolitical environment.

With all these factors in mind, LDP kingmaker Shin Kanemaru's attempt on diplomatic normalization with the DPRK in 1990 was expected to include Japan's final apology for thirty-six years of occupation in Asia and to terminate the forty-five years of abnormal relations with North Korea that had pertained since World War II. Once normalization had been brought about, the Japanese government would have to pay billions of dollars in reparations, part of which might have been contributed by the diaspora community through the pachinko industry.[62] Such a scenario, however, readily collapsed due to changes in Japanese domestic politics. After Kanemaru fell from power and was arrested for tax evasion in the Sagawa Kyubin scandal in 1992, no one took on his role in managing the relationship between the *Chongryun* and North Korea. It quickly came to be seen as a political liability and any interested parties backed away from the responsibility of managing this relationship. In addition, the Japanese press coverage of the North Korean kidnapping of Japanese civilians shocked the Japanese nation[63], resulting in domestic pressure on the government, while Pyongyang refused to cooperate on the issue. Moreover, when the Japanese economy collapsed in 1991, the *Chongryun* pipeline to both Pyongyang and the domestic Japanese political parties began to shrink as bankruptcies spread through the pachinko industry and a new generation of North Koreans born in Japan lost their ideological fervour for Pyongyang. According to Nicholas Eberstadt, *Chongryun* remittances never surpassed an annual figure of $100 million after the collapse of the bubble.[64] The direct North Korean threat to Japan began to grow under these conditions.

The North Korean nuclear crisis of 1993–4 shifted attention back to Korea and regional security, which inevitably allowed increased cooperation between existing allies. This cooperation established a strategic convergence between the two north-east Asian alliances, strengthened the Japan–US and ROK–US alliances against the North Korean threat and re-aligned priorities to focus primarily on the nuclear issue.[65] Alliance cohesion again shifted into a position of strength under asymmetric alliance in response to the external threat posed by the North Korean nuclear crisis. We can then test this hypothesis: if alliance bloc cohesion amongst Japan, South Korea and the US is strong, then Japan's policy will be one of exclusion toward a diaspora group supported by an enemy homeland, but one of inclusion to a diaspora group supported by an allied homeland (**Configuration 1**).

First of all, in terms of policy towards a pro-North Korean diaspora, after launching a Nodong missile in 1993, the Japanese government quickly received instructions from the US to cut off *Chongryun* remittances to North Korea, to close financial institutions operated by *Chongryun* and to prevent the pro-Pyongyang diaspora from travelling to North Korea. Although Washington expected much more, the National Police and Finance Ministry reluctantly followed US instructions, but warned that these steps would not stop the entire flow of money to Pyongyang, and would lead to violent protests from the North Korean community in Japan.[66] Tokyo still maintained its own unofficial link with North. In March 1995, just after the KEDO project was launched, the three political parties in Japan's coalition government (the LDP, SDPJ, and Shinto Sakigake) signed an agreement in Pyongyang with the Korean Workers Party calling for a resumption of normalization talks. However, Tokyo agreed to proceed more slowly with its talks with North Korea after South Korean President Kim Young Sam expressed concern in a summit with Hashimoto in January 1997 that a new round of Japan–DPRK talks might complicate the Four Party talks.[67] As a response to this cooling-off on Japan's part, North Korea launched a three staged Taepo-dong ballistic missile directly over Japanese airspace on 31 August 1998. North Korea called it a satellite launch, but the event was a shock not only for the Japanese government but also for the public, and led Tokyo to protest for the first time by linking the nuclear issue with the missile issue.[68] Facing severe criticism at home, Japan unilaterally suspended its negotiations with the DPRK in Beijing and announced that it would suspend all food aid for the North and financial support for KEDO.

Under heavy pressure from Washington and Seoul, the Japanese government eventually signed the KEDO cost-sharing agreement on 10 November 1998, pledging to contribute US$1 billion to the LWR project.[69] Angry LDP politicians, however, called for harsher responses toward the North and for the first time the Party's Policy Affairs Research Council began investigating measures to cut off *Chongryun* remittances to the North.[70] The Taepo-dong launch allowed Japanese host state elites to admit the real possibility of a direct attack on Japan by North Korea. The North Korean community therefore became the target of security measures in the aftermath of the missile test.

After the Taepo-dong launch, the Japanese government discovered a new enthusiasm for cooperation with Seoul on matters of security, which influenced policymaking toward the pro-South Korean diaspora into a more accommodating direction. In 1998, the South Korean President, Kim Dae Jung, met with Japanese Prime Minister Keizo Obuchi, reaching an agreement to open up a cultural exchange by the establishment of a Korea–Japan partnership, which until then Korea had prohibited. This exchange served to promote the Korean entertainment industry in Japan, and from that point on, Japanese acceptance of Korean influence and culture Korea improved greatly. South Korea suddenly became closely allied to the Zainichi in cultural terms, which stemmed from the *Hallyu* boom in Japan.[71] In contrast, supporters of North Korea began to dwindle as the media's coverage on the negative socio-political situation in the North became more extensive. An increasing number of previously pro-North Koreans and their subsequent generations who had not registered as South Koreans in the normalization treaty in 1965 now chose to acquire South Korean nationality. Rather than focusing on their divided homeland's political issues, the Zainichi Korean diaspora reduced the degree of their activism in the post-Cold War era.[72] Then, more concerned about their life in Japan than any homeland-oriented issue, Zainichi social movements demanded the right to participate in local elections, which became more prevalent, particularly among those affiliated with *Mindan* (the pro-South Korean association). Zainichi social movement activists lobbied for voting rights in local assemblies, pointing out that they paid taxes and, as such, their voices deserved to be heard in the community. As a counter-movement to the foreigners' voting rights, however, in 2001 some conservative politicians participated in preparing a bill to grant special permanent residents (mostly Zainichi Koreans) the right to acquire Japanese nationality.[73] The bill on foreigners' voting rights was then submitted to the 2001 National Diet. This was defined as an attempt to acquire fully fledged national citizenship, as opposed to the possibilities of diasporic citizenship.

What happens to different types of citizenship when we attempt to accommodate group differences and try to articulate broader principles at the same time? First, as the legal status of resident Koreans was stabilized in the mid-1990s and unified into the same post-colonial identity, civil rights activists shifted their focus to foreign residents' local voting rights. The acquisition of such rights would emphasize Zainichi identity as it would expand 'local citizenship' – the granting by local governments and organizations of basic socio-political rights and services to immigrants as legitimate members of these local communities.[74] Local citizenship acquisition was not incompatible with homeland-oriented identity politics, and the South Korean government also began to improve the rights of overseas Koreans through a series of legal measures. In 1999, South Korea passed the Overseas Korean Act that created an Overseas Korea (F-4) visa categgory, which gave eligible co-ethnic immigrants access to health insurance, pensions, property rights, unrestricted economic activity and broader employment opportunities.[75]

The counter-move proposing easy access to naturalization was supported by some Zainichi Korean civil rights activists to reflect the concept of Korean–Japanese citizenship, although their goal was not simply a host-oriented direction. While discourse on acquiring Japanese nationality already existed, it was the acquisition of Japanese nationality through via assimilation that was now being discussed. What was different from previous discourses was a proposal of Korean–Japanese identity that indicated status as an ethnic minority member of a multi-ethnic Japanese nation. For example, Young-wi Kou, a Zainichi Korean lawyer, established LAZAK with other Korean lawyers in 2001. Kou claims that his position differed from the existing naturalization system, in which Japanese nationality is granted by permission from the Justice Minister. He asserts that Zainichi should participate in democracy as Japanese nationals while maintaining Korean ethnicity through education and by continuing to address postcolonial problems.[76] Although Kou stressed the importance of gaining the right to participate in national politics, his concept was much closer to the idea of transnational identity.

Interestingly, an affinity exists between citizenship and identity insofar as they are both group markers. Citizenship demarcates the members of one polity from another, as well as separating members of a polity from non-members. Identity differentiates one group from another as well as allowing for the possibility that groups could be targets for assistance, hatred, animosity, sympathy or allegiance. The difference between citizenship and identity is that the former carries legal weight while the latter carries social and cultural weight.[77] Whilst it is true that a tension exists between citizenship and identity, by expanding the scope of citizenship via the recognition of multiple group rights and identities, post-modernization and globalization compel us to move beyond essentialist and constructivist assumptions of identity. The earlier stage of the post-Cold War era, 1993–2002, did not only open a new geopolitical space but also compelled the Zainichi diaspora to re-think their sense of belonging to either North or South Korea or Japan, and to move on to the acquisition of nationality by associating naturalization with a new identity category as Korean–Japanese. Indeed, the years 1993–2002 witnessed a strong alliance cohesion due to the North Korean nuclear crisis, which supports **Configuration 1**: if alliance bloc cohesion is strong, the host state policy will be one of exclusion toward a diaspora group supported by an enemy state, but one of inclusion toward a diaspora group supported by an allied homeland. As far as the North Korean diaspora was concerned, the Japanese government started to investigate how to cut off *Chongryun's* remittances to North Korea.

Coinciding with the Japanese government's exclusionary policy in response to the North Korean nuclear crisis, Japanese public opinion and media also reacted against the pro-North Korean diaspora in the same direction. The negative view of North Korea as a 'rogue state' inevitably impacted the everyday life of the Zainichi. Korean children attending *Chongryun* schools, for example, faced many incidents of harassment. Since North Korea triggered a controversial public debate regarding its development of nuclear weapons, many Korean schoolgirls became targets of humiliation and even physical violation from the Japanese public because their traditional Korean *chima chogori* costume made them easily identifiable with North Korea. There were reports of female students having their school uniforms cut off them. In 1994, when North Korea's nuclear weapons programme was increasingly attracting increased Japanese media attention, a total of 154 incidents of abuse toward Korean schools were reported between April and July alone.[78] An increasing number of North Korean diaspora members chose to acquire South Korean nationality, thus offering them the possibility of naturalization,[79] while the government became more lenient in the mid-1990s, granting Japanese nationality to resident Koreans who possessed South Korean nationality, allowing about 10,000 Koreans per year to become nationals.

Alliance Constraints: The Rise of China and the Division on North Korea Policy (autumn of 2002–14)

The post-Cold War years of 1990–2002 saw a transition in the Japan–US alliance from a time of separation (1990–2) into one of consolidation and the deepening of security ties with South Korea, the so-called virtual alliance (1993–2002). However, in the years from autumn 2002 until the present, the consensus of these two strategic alliances began to grow increasingly constrained in its management of relations with the Korean peninsula. The following sections highlight areas examining the level of external threat, military capabilities, shared policy goals and the availability of substitute allies; they examine the extent of alliance cohesion and whether or not alliance cohesion still matters in terms of policymaking towards the Zainichi diaspora in the great power reconfiguration of the post-Cold War years. By identifying additional factors relating to the limitation of correlation between alliance cohesion and the diaspora as a category, the section concludes by examining the conditions under which a host state treats a diaspora, either inclusively or exclusively, in reflection to the shift of alliance cohesion.

Assessment of the Value of Alliance and the Availability of Substitutes
During the Cold War, economic and security benefits flowed in the same direction within political blocs and trade between adversaries was negligible. But since the end of the Cold War, trade and investment patterns have developed with fewer political constraints, enabling broadened economic opportunities for doing business with former enemies.[80] China's emergence as South Korea's leading trade partner (surpassing the United States) reinforces questions about the durability of the US–South Korean alliance. In 2003, China overtook the United States as South Korea's top export market. In 2004, China replaced the United States as South Korea's top trading partner.[81] China's rise puts South Korea into a strategic dilemma as far as the United States and China are concerned. South Korea has to manage its security – which is grounded in the ROK–US alliance, and its economic well-being – which is dependent on the ROK–China strategic cooperation partnership. China influences the Korean peninsula in both economic and political terms, a fact that allows South Korea to recognize its image as a reliable partner to an even greater extent. Given that 28 per cent of Pyongyang's trade was conducted with Beijing, and China's provision of food and energy for North Korea was on highly preferential terms, no economic sanctions against the North could be fully implemented without Beijing's cooperation.[82] China thus emerged as a crucial mediator for the region, significantly boosting its image as a responsible great power. To address the North Korean nuclear crisis, China hosted a trilateral consultation with North Korea and the United States in April 2003 and subsequently sponsored the six-party talks among those three nations plus South Korea, Japan, and Russia in August of the same year.[83]

The extent of China's economic and political influence on the Korean peninsula poses challenges for Japan as far as security cooperation with South Korea in the context of the US–Japan security alliance is concerned.[84] The South Korean government has had a quasi-alliance with Japan by virtue of the fact that both South Korea and Japan are allies of the United States. During the Cold War, Japan played a role in the security sphere as a base through which US forces would flow in order to ensure security in the event of a threat to South Korea from the North.[85] China's constructive role in managing the North Korean nuclear problem has gradually allowed a large number of South Koreans – including diplomats, politicians and policy analysts – to endorse the belief that South Korea's diplomacy should break out of its heavily exclusive reliance on the United States.[86]

The Korean peninsula has historically been an object of past competition over influence and control between China and Japan. This competition led to military conflict in the mid-1890s, and the eventual annexation of Korea in 1910. As the US Cold War strategy logic fell into decline after the Cold War ended, a growing gap in perception between the United States and South Korea emerged regarding Japan, a gap which was accompanied by historical memory. Although Washington has been assigning an increasingly expanded military and strategic role to Tokyo, from South Korea's viewpoint, with the fourth-most-powerful navy and third-largest defence spending in the world, Japan is viewed as the nation that poses the biggest threat to South Korea's security interests.[87] While South Korea has managed to sustain an amicable official relationship with Japan under the auspices of the United States, security concerns have always lurked in the background.[88] These perceptions are shared between China and South Korea as well.[89]

The ties between South Korea and China have become an important cornerstone of regional cooperation in East Asia. China has become a crucial intervening variable in South Korea–US relations, while simultaneously becoming part of a strategy to strengthen relations with all the regional powers to promote inter-Korean cooperation. The strongest advocator of this policy was South Korea President, Roh Moo-hyun (2003–8) who began to mediate between China and the United States and to position South Korea between Japan and China, where it could be seen as the 'Northeast Asia balancer.'[90] According to Roh and his advisors, the US–Japan alliance and the US–ROK–Japan triangular alliance were viewed as relics of the past that now served only to antagonize relations with North Korea and China. They perceived Asia as being divided between South Korea, the US and Japan, a southern tripartite alliance that traditionally stood against North Korea, China and Russia, the corresponding northern tripartite alliance, and advocated that South Korea, while continuing the US–ROK alliance, should not be confined to the limitations of that southern alliance.[91] This presented a major challenge to the strategic position of Japan and the Japan–US alliance vis-à-vis the Korean peninsula and Northeast Asia.

Assessment of the Value of Alliance: Shared Goals

The 9/11 terrorist attack on the United States and the 2003 Iraq War shifted Washington's strategic focus from regional threats to the global war on terror. On 29 January 2002, George W. Bush classed North Korea, together with Iraq and Iran, as part of 'an axis of evil' in his State of the Union address. In this speech, Bush claimed that North Korea's crime was that it was a regime arming itself with missiles and weapons of mass destruction, while starving its own citizens. Bush considered that North Korea was in violation of two significant international norms; non-use of nuclear weapons and human rights concerns. Indeed, since 9/11, the North Korea factor had become an independent variable in terms of threat, one that represented not only a regional but also a global challenge to peace and stability. This has made alliance management a great deal more difficult. Alliance requires the coordination of foreign policies, meaning that one or both allies must abandon policies supporting individual interests that might separate the allies in order to achieve the shared goals of peace and stability.

Japan and the United States continued to cooperate with South Korea on nuclear non-proliferation on the Korean peninsula. This served a global purpose as well as a regional one, but after September 2002, when Prime Minister Junichiro Koizumi made his historic visit to Pyongyang and made a breakthrough in normalization talks with North Korea, the allies could no longer focus solely on nuclear non-proliferation. Another globally accepted issue, that of human rights, became controversial because the North Korean leader, Kim Jong Il, admitted, for the first time, North Korean involvement in the Japanese abductees issue. The DPRK promised to take appropriate measures so that these regrettable incidents would never recur in the future. For its part, the Japanese government demanded investigations into thirteen abducted Japanese subjects, and the North Korean authorities responded by admitting that four were alive, eight had died, and one had never entered North Korea.[92] Human rights issues had become widely accepted in Japan, and the abductee issue invoked a rise in Japanese neo-nationalism, which culminated in an explosion of an Anti-Korean sentiment in the wake of Kim Jong Il's admissions. Since that time, handling the abductees issue has become a major diplomatic challenge for Japan and the United States.[93]

In order to balance policy priorities towards North Korea during the period from 2002 to 2006, Koizumi and Bush strove to support policy alignment between the US and Japan, and maintained close cooperation in both the nuclear talks and the abductee issue. In the wake of the second North Korean nuclear crisis in October 2002, Japan collaborated with South Korea and pressured the United States towards the establishment of the Six-Party Talks that included China and Russia, in order to work out a nuclear deal with North Korea from 2003.[94] The Bush administration supported Koizumi behind the scenes, since the issue of human rights violations was high on Bush's own global agenda. The US administration conducted its own human rights campaign on North Korea, passing the North Korea Human Rights Act in October 2004 and inviting the parents of one of North Korea's victims, Megumi Yokota, to testify in a Congressional hearing in April 2006.[95] However, with presidential elections looming the Bush administration was running out of time, and began to place a higher priority on forging a nuclear deal. Japan–US cooperation began to falter as the abductees issue took second place to the nuclear issue in the Six-Party Talks, and the two countries ended up on a collision course in 2007–08. During the initial actions in readiness for the Implementation of the Joint Statement in February 2007, when the Bush administration removed North Korea from its list of terrorist-supporting states to make better progress on the nuclear issue, Japan's position was that North Korea should be removed only after it had released Japanese abductees and demonstrated willingness to disable its nuclear programme. President Bush, however, prioritized the nuclear deal and finalized the removal of North Korea from the list of state sponsors of terrorism on 11 October 2008, upon agreement to a verification framework for North Korea's nuclear sites.[96] Despite the climbdown over the abductees, the Bush administration still failed to barter a satisfactory nuclear deal. Both issues, nuclear non-proliferation and the Japanese abductees, still remain unresolved.[97]

The gradual drift between Japan and the United States also extended to the relationship between Japan and South Korea. From South Korea's point of view, Japan inflated the importance of the abduction and missile issues, which did more harm than good as far as trilateral consultations with the United States and Japan were concerned. The United States found itself sandwiched between two conflicting allies, and had a difficult time coordinating its efforts regarding the two countries.[98] Throughout the Six-Party Talks, it had become obvious that because of the territorial disputes and historical issues that were constantly sizzling between them, relations between Japan and South Korea had grown particularly tense. When the territorial dispute over Takeshima/Dokdo Island emerged in the spring of 2005, the South Korean delegation refused to sit at the same table as the Japanese delegation.[99] The current South Korean President, Park Geun-hye, who was sworn in as the head of state in February 2013, refused to hold summit with Japanese Prime Minister Shinzo Abe, citing his own misperception of history and the lack of repentance over past brutalities.[100] However, US President Barack Obama finally brought Abe and Park Geun-hye together at the nuclear summit in The Hague in March 2014. 'Close co-ordination between our three countries has succeeded in changing the game with North Korea,' President Obama said, and continued:

> Obviously Japan and the Republic of Korea are two of our closest allies in the world and our two most significant and powerful allies in the Asia Pacific region. The ties between our peoples run deep. We do an extraordinary amount of trade together. Our alliances with South Korea and Japan uphold regional peace and security. So our meeting today is a reflection of the United States' critical role in the Asia Pacific region, but that role depends on the strength of our alliances.[101]

Although bilateral consultation between Washington and Seoul or Tokyo had never been problematic, it has become extremely difficult to conduct trilateral consultation between the United States, Japan and South Korea since the beginning of the Six-Party Talks. The lesson that can be learned from this is that if allied states find themselves disagreeing in deciding priorities for achieving a desired goal, they must coordinate with each other even more strongly to eliminate conflicts of interest that could separate them.[102] Credible alliances allow individual states to enforce national security and accomplish foreign policies more effectively and efficiently – yet alliances come with costs as well, and states must accept such continuing and potential costs and make compromises to preserve alliances.[103] These costs include managing the alliance, and increased foreign policy consultation and coordination are essential to manage a credible joint position. Potential costs are those which leaders should expect to incur should the full obligations of the alliance be invoked.[104] Alliances signal to external parties the willingness of allies to come to one another's aid if threatened by outside nations, and the strength of an alliance is measured by the degree of coordination that exists within its membership. Tighter allies fight more effectively together, but their greater coordination imposes higher peacetime costs.[105] In fact, the Tokyo government based its participation in the US-led security operations in Iraq partly on the belief that the US would reciprocate by joining Japan in trying to resolve the North Korean threat, and emphasized the need to maintain the Japan–US alliance.[106] Koizumi's trip to North Korea in 2002 tested the degree to which the US–Japan alliance signalled its solidity concerning the Korean peninsula to parties outside the alliance, and showed that this coordination had come at a severe and on-going cost.

Assessment of the Value of Alliance: Threat Level
From the autumn of 2002 to the present, Japan has felt more vulnerable in the defence of its nation due to heightened concerns about North Korea and China. The most overt threat to the US and Japan is the proliferation of ballistic and Cruise missiles, weapons of mass destruction which can strike American and allied interests throughout the region (see Figure 5.1). The North Korean contribution to this threat is well-known. North Korea has developed and tested nuclear weapons and an array of ballistic missiles that can reach targets throughout Asia. The danger is heightened by the instability of the North Korean regime, which poses a direct threat to the security of Asia and the world. Given North Korea's new military capabilities, US and Japanese policymakers have been forced to re-evaluate what

North Korean missile ranges
Maximum estimated/calculated

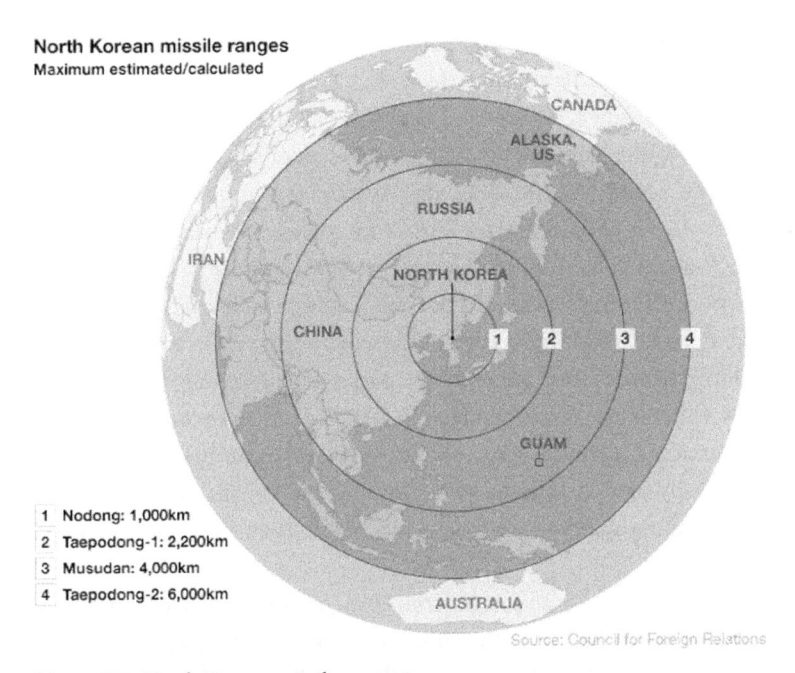

1 Nodong: 1,000km
2 Taepodong-1: 2,200km
3 Musudan: 4,000km
4 Taepodong-2: 6,000km

Source: Council for Foreign Relations

Figure 5.1 North Korean missile ranges[a]
[a] BBC World News, 9 July 2014, available from: http://www.bbc.co.uk/news/world-asia-28223183 (accessed 9 July 2014).

it might mean even to consider a renewed Korean War. In the event of a North Korean attack on Japan, for example, Washington and Tokyo would be forced to figure out from scratch what the two sides may expect from one another with very little time in which to mount a credible response.[107] Such anxieties have created concerns about the reliability of the US alliance, centring on America's defence commitment and its capability for defending Japan and the maintenance of regional stability – especially on the Korean peninsula in the wake of North Korea's first nuclear test in October 2006. It has also become increasingly clear that Japan is unable to fulfil its role as America's partner not only in global and regional security operations, but also regarding Japan's own defence due to constraints in the Japanese constitution concerning the use of force and collective self-defence, which were endangering the alliance.[108]

There is also another threat facing the alliance – the growth of Chinese military power. In addition to the growing missile menace, the United

States and Japan face an unprecedented threat to their traditional dominance of Asian–Pacific airspace. Recently, China has fielded large numbers of Su-30MKK attack aircraft and begun serial production of the indigenous J-10 fighter, announcing Beijing's emergence as an aerospace power in its own right.[109] In late 2007, the Japanese government reported that Chinese Air Force fighters were repeatedly approaching Japanese territory on high-speed runs, leading Japanese defence officials to conclude that these were drills to prepare for a possible Cruise missile attack on Japan.[110] In the sphere of maritime security, US analysts also believe that China has developed its submarine forces for the purpose of preventing US and allied access to the western Pacific and to threaten vital sea lanes such as the Strait of Malacca.[111] China's growing submarine presence puts pressure on maritime trading nations to avoid offending Beijing, and increasingly jeopardizes the resources and trade flows upon which Japan and other American allies depend.[112]

China's aggressive action regarding its dispute with Japan over the Senkaku/Diaoyu Island in the East China Sea is a typical example for China's growing submarine presence. Tension has increased there due in part to the frequent penetration of the 12 nautical mile territorial sea cordon around the islands by China's civilian patrol vessels.[113] Regional tensions expanded widely after a Vietnamese fishing boat sank on 26 May 2014, following a collision with a Chinese ship. The incident occurred after China had deployed an oil drilling rig in disputed waters around the Paracel Islands in the South China Sea.[114] At the annual Shangri-La Dialogue in Singapore,[115] during the 30 May–1 June dialogue, US Defense Secretary Chuck Hagel scolded China for undertaking 'destabilizing unilateral actions' and in this context he added that 'we oppose any nation's use of intimidation, coercion or the threat of force to assert their claims.[116]' Hagel's words also included a warning that the US will 'not tolerate any attempt to alter the status quo by force or coercion.' The concepts of the status quo and international order exist, of course, in the eyes of the beholder. Used in a regional sense, the US apparently uses these terms to refer to any situation in which it is the dominant actor and patron. To the US, the status quo is essentially a continuation of its Cold War policy and posture in the region – a substantially forward-deployed military presence under a hub-and-spoke alliance structure. However, the status quo in the region itself is certainly changing. As such, whilst the US may consider itself the sole superpower in the present as well as the past, its credibility, legitimacy and ability to impose its will are gradually fading. Moreover, Japan is now trying to regain its nation state status, with its own full powers of defence, while re-developing its militaristic past in response to external threats such as the North Korean nuclear crisis and the growth of China.

Assessment of the Value of Alliance: Military Capability
Shinzo Abe's administration (2012–present) is apparently no longer interested in planning any further diplomatic strategy aimed at stabilizing Japan's relationship with China. On 17 December 2013, the Cabinet approved new National Defence Programme Guidelines and a Mid-Term Defence Programme, which are in line with the country's new strategy. The proposal to allow Japanese SDF to strike enemy missile bases has alarmed Japan's neighbors and the United States.[117] The new National Defence Programme Guidelines say only that Japan will 'consider' how to deal with ballistic missile attacks and 'take necessary measures.' The strategy also indicates the need to sanction the right of collective self-defence. One of the document's goals is 'heightening the effectiveness of the Japan–US national security framework and realizing a multifaceted Japan–US alliance.'[118] Specifically, Abe wants to rewrite the government's decades-old interpretation of the pacifist Constitution to lift Japan's self-imposed ban on exercising the right of collective self-defence. Such a reinterpretation would allow Japan to launch counterstrikes should one of its allies come under attack, and could pave the way for the SDF's participation in multinational actions.[119] Japan's national security strategy addressed concerns about China's intrusion into Japanese territorial waters near the Senkaku Islands in the East China Sea as well as its unilateral setting of an air defence identification zone in that region. The military posturing by China and North Korea is also being used as justification for strengthening the SDF.

Under the new National Defence Programme Guidelines, the upper limit for GSDF members has been set at 159,000; 5,000 more than the previous guidelines, which were compiled when the Democratic Party of Japan was in power. The Mid-Term Defence Programme has set total defence spending over a five-year period at 24.67 trillion yen ($240 billion), an increase of about 1 trillion yen over the previous defence programme.[120] Abe is trying to bolster Japan's defence capabilities as part of his plan to broadly remodel the nation's security apparatus in the light of the fact that China is starting to make its presence felt and North Korea's missile and nuclear development programmes seem to be escalating. At the same time, he is strengthening bilateral ties not just with Japan's top security ally, the US, but also with Southeast Asian countries and Australia to counterbalance the rise of China.[121] Pentagon chief Chuck Hagel told the forum that the US will support Abe's efforts to 'reorient its collective self-defense posture toward actively helping build a peaceful and resilient regional order.'[122]

Throughout this process, however, the absence of sustained, senior-level trilateral dialogue between Washington, Tokyo and Seoul has prevented the growth of coordinated capabilities. Although American policymakers would welcome such capabilities, Seoul and Tokyo are watching each other's military capability development with mistrust.[123] The net result is that while the United States is achieving its immediate goals, ongoing antagonism between Seoul and Tokyo threatens the long-term health of both alliances. Now another possible trilateral collaborative partner is emerging. Australia is becoming the third link in an additional trilateral security dialogue with the United States and Japan, while Shinzo Abe's government is pursuing quadrilateral dialogue with the United States, Australia and India, which provides yet another example of the need for broader discussion among the Asian powers that share both strategic interests and liberal values.[124]

Summary of the Assessments on Alliance Cohesion (2002–14)
The US–Japan–ROK trilateral alliance has been challenged in the years between 2002 and 2014 by strategic changes such as the rise of China and North Korea's nuclear and missile development. Washington has sought to improve cooperation, but historical animosities between Seoul and Tokyo have impeded these efforts. After the earlier years of the post-Cold War period, 1993–2002, the North Korean threat has re-focused US–Japan–ROK priorities into a state of virtual alliance. Since the autumn of 2002 and Koizumi's trip to North Korea, the strategic convergence on the North Korean threat between US allies South Korea and Japan has been negatively influenced by the issue of Japanese abductees and the North Korean

nuclear crisis. Japan and China are locked into a more serious confrontation in the East China Sea, where China has been challenging Japanese sovereignty and control over the uninhabited Senkaku/Diaoyu Islands. In other words, the US–Japan–ROK alliance, despite its achievements of solidarity in the 1990s (more specifically between 1993 and 2002), stands challenged by strategic changes that have occurred between 2002 and the present, and the period has witnessed weak alliance cohesion between the US, Japan and the ROK. The North Korean provocation and threats continue, but it has become apparent that they would not be met by a united response from the allies. The most striking strategic changes that have broken the consensus which continued from the Cold War years into the 1990s are the rise of China, the decline of Japan and the weakening of American hegemony. These, as well as changes in the geopolitical balance of power, have allowed US strategic satellite allies Japan and South Korea to re-consider their own national strategies, which has inevitably affected Japanese policies towards the Zainichi diaspora as they belong to both countries as a host country and homeland.

Alliance Cohesion and the Diaspora in the Post-Cold War Era (2002–14)

During the Cold War era, weak alliance cohesion allowed Japan to expand its autonomy by using the diaspora card informally, encouraging Japan to accommodate a diaspora group which was supported by an enemy country, North Korea. The ambivalent attitude of the Japanese government toward the North Korean regime was driven by Japan's two-Korea policy, which allowed it stay neutral because of security and economic motivations. Rather than referring to the sensitive two-Korea policy, more relevance was officially ascribed to the policy of *Seikei Bunri* (the separation of economics from politics), under which Japan officially recognizes the ROK but not the DPRK whilst continuing to trade with both. This two-Korea policy has inevitably affected Japan's treatment of Korean nationals. With no official normalization between Japan and North Korea, the *Chongryun* headquarters was used as a *de facto* embassy for North Korea, allowing Japan to promote an accommodation policy by maintaining the cultural specificities of the diaspora group.

The years since 2002 have witnessed weak alliance cohesion, yet, unlike the Cold War era, Japan's policy towards the Zainichi diaspora group supported by North Korea has been exclusive, even under the conditions of weak alliance cohesion. The deterioration in Japan–DPRK relations since the autumn of 2002 is increasingly putting *Chongryun* under pressure from the Japanese government. The pressure dates back to 17 September 2002, when Japanese Prime Minister Junichiro Koizumi visited Kim Jong Il in Pyongyang in order to work towards normalization by settling the issue of nuclear arms development in the DPRK and obtaining clarification regarding the thirteen Japanese citizens who were suspected to have been abducted by North Korean secret agents. Kim Jong Il's personal admission to the abductions was surprising, and he reported that eight of the abductees had died under unclear circumstances. To make matters worse, North Korea admitted in October 2002 that it had secretly resumed a nuclear weapons programme, in violation of the 1994 nuclear non-proliferation treaty signed with the United States. North Korea expelled UN inspectors, severely worsening relations with Japan. Indeed, the deterioration of DPRK–Japan relations has provided the Japanese government with a reason to crack down on some of *Chongryun*'s activities.

In June 2003, after Koizumi's trip to North Korea, the Japanese authorities blocked the departure of the *Chongryun*-operated ferry *Mangyongbong-92*, which made a trip to North Korea about every two weeks, claiming that the ship was being used to smuggle money and military technology to North Korea. Additionally, the Japanese Justice Ministry pressured banks to suspend remittances to North Korea, and the Ministry of Education refused to accredit Korean schools. These factors, which explain the contradictory reaction to the North Korean threat, were further complicated by two nuclear missile launches: the 1993 Nodong and the 1998 Taepodong missile tests. During the earlier post-Cold War years between 1990 and 2002, even though the United States often required Japanese help to cut off *Chongryun*'s remittances to North Korea or to close its financial institutions to prevent Diaspora funds travelling to the North, Japan was reluctant to do so, warning that these steps would not stop all money flowing to Pyongyang and would lead to violent protests from the North Korean community in Japan.[125] When conditions changed from the Cold War to post-Cold War circumstances, the host state treated the Zainichi in a different way to that suggested by the hypothesis. This treatment depended on two additional factors relating to the restriction of the relationship between alliance cohesion and the diaspora as a category.

The two factors in question were the rise of China and the North Korean threat, which ultimately related to security concerns. As explained in the previous chapter, during the Cold War era, the two Korea policy (or *seikei bunri*) automatically related to Japan's policy towards China, that is, the One-China policy. The One-China policy meant recognizing one sole legitimate PRC government, and seeing Taiwan as part of its territory, a vision which was demanded in the September 1972 at the Tanaka–Zhou summit when Japanese Prime Minister Tanaka met with Chinese premier Zhou Enlai to seek a normalization treaty between the two countries. Japan's relations in East Asia were primarily determined by the conjunction of mercantile interests and past US strategies. During the Cold War era, Japan could accept China's demands for a One-China policy because of its confidence that Japanese economic leadership would integrate China on Japan's terms, but only under the condition of weak alliance cohesion. In the post-Cold War era, such confidence has ebbed, and a new geopolitical configuration has emerged due to the rising power of China in comparison to the decline in Japanese economic influence. In the post-Cold War era, especially since 2002, Japan's space for manoeuvre has been narrowed down by using the diaspora card as an instrument to expand Japan's own autonomy in order to stretch in space to cross state boundaries. Japan no longer needs to keep ties with North Korea in order for China to move in favourable directions or to sustain Japanese economic growth.

The North Korean threat was no longer localized to the Korean peninsula (a Korean contingency) but became an immediate threat to the Japanese homeland (a Japanese contingency).[126] Although in the earlier stages of the post-Cold War era between 1990 and 2002, the North Korean threat became the catalyst for closer Japan–US and US–South Korea defence cooperation, it has become more complex since Koizumi's trip to North Korea in September 2002. The Japan–US and Japan–ROK alliances on the North Korean threat prioritized focus primarily on the nuclear issue, while other issues such as missiles and abductees had to be dealt with separately.[127] After the end of Cold War, the overall national and military balance of power shifted in favour of South Korea, making made full-scale war difficult for the North. As China's power increases, South Korea has become closer to China over recent years as opposed to its more distant stance during the earlier years of the post-Cold War era (1990–2002). During the Cold War, the alliance cohesion between Japan and South Korea had always been connected to America's defence commitment on the Korean peninsula. At that time, Japanese politicians often attempted to manoeuvre deftly between the two Koreas without abandoning Japan's *de facto* alignment with the South and to remain neutral within East Asia for as long as Japan saw a strong enough security guarantee from the United States.[128] Now, in the post-Cold War years since 2002, South Korea has nurtured a new relationship with China and plays a more influential role as a middle power in East Asia.[129]

Thus, during the period since 2002, Japan has felt more vulnerable in defence due to heightened concerns about North Korea and China. These anxieties have allowed Japan to react directly in response to external threat. In the later post-Cold War years, Japan's policymaking towards the Zainichi diaspora is exactly applicable to Mylonas' nation-building theory. Mylonas pointed out that if the host state has an ethnic group which is supported by kin from an external enemy state, the host state is not likely to initiate a policy of maintaining the cultural individuality of that group or supporting its minority status by supporting that cultural individuality or its minority status by institutionalizing it, which means that a stricter policy will be enforced – one of exclusion. Among the causal pathologies that affect policy choices, Mylonas considers that a host state is likely to exclude a non-core group if that host state has revisionist aims and an enemy is supporting the non-core group. As a matter of fact, it was Junichiro Koizumi, Japanese Prime Minister from 2001 to 2006, who expended his greatest energies on the strengthening of the US–Japan alliance by providing support for the United States in Afghanistan and Iraq. Koizumi widened the operational scope of JSDFs, which had been limited by the Constitution Article 9, by enacting special laws to dispatch JSDFs to those combat areas.[130] Also, in order to counter the rise of China, Koizumi proposed that Australia and New Zealand join a regional framework in East Asia as core members.[131] Indeed, since the Koizumi administration, Japan's foreign policy has become proactive, working in a revisionist direction.

Japanese Prime Minister Shinzo Abe, in the first year of his second term in office in 2013, did not hesitate to reveal his posture as a historical revisionist, notably with respect to wartime Japan.[132] Following his predecessor Koizumi, his controversial visit to the Yasukuni shrine, which memorializes the war dead, including Class A war criminals such as Hideki Tojo, shows that he is hoping to engender self-confidence and patriotism within the Japanese public by encouraging a spirit of nationalism. Abe is attempting to reinterpret the constitution to allow for the exercise of the right to collective self-defence.[133] Abe will also formally abolish Japan's decades-old ban on weapons exports. In January 2014, his administration revised textbook screening guidelines to give Japanese children a more patriotic take on modern Japanese history and to better reflect the government's view on territorial issues such as the Senkaku Islands.[134]

Therefore, the post-Cold War years of 2002–14 exactly match the causal pathway presented by Mylonas; that a host state is likely to exclude a non-core group when the state has revisionist aims and an enemy is supporting the non-core group. After Kim Jong Il's admission regarding the abductions in 2002 and the subsequent North Korean nuclear crisis,[135] the Japanese government undermined *Chongryun*, which led to its eventual dissolution. *Chongryun's* role was to promote ethnic attachment to North Korea firstly through the administration of Korean ethnic schools, secondly via repatriated family members in North Korea, thirdly by official exchange of gifts, letters, and financial assistance with the North Korean government and fourthly by the political participation of elite association members in North Korea's politics.[136]

To promote the ethnic attachment of the Zainichi Koreans to North Korea, *Chongryun* built numerous Korean schools and urged its members to send their children to them. In 2005, *Chongryun* sponsored 45 kindergartens, 62 elementary schools, 38 middle schools, 11 high schools, and one university.[137] Its Central Education Institute works closely with the North Korean government to oversee the direction of Korean Nationalism of and against Zainichi Koreans in Japan. Until recently, portraits of Kim Il Sung and Kim Jong Il hung in front of most classrooms and inside all university dormitories. The North Korean government encourages *Chongryun* members to study the Korean language and history at Korean schools in Japan as part of its 'loyalty education subjects.' The North Korean government has historically supplied these schools with textbooks from North Korea designed specifically for Korean children living in Japan. It sent more than 45 billion yen (US $413 million) to support Korean education in Japan between 1957 and 2005.[138]

Mindan also operates four Korean schools in Japan: the Tokyo Korean School, the Kyoto Korean School, the Osaka Keum Kang Institute, and the Osaka Keon Kook School. Unlike *Chongryun* schools, however, most students in *Mindan* schools are not the children of Zainichi Koreans. The Tokyo Korean School, for instance, teaches mostly the children of South Korean diplomats and businessmen who are temporarily living in Japan.[139] The post-Cold War era years since 2002, however, have allowed the Japanese government to issue a blatant act of discriminative ordinance between *Chongryun* schools and *Mindan* schools. In February 2003, the Japanese Ministry of Education issued an edict giving all foreign high schools including *Mindan* schools full accreditation, while refusing to regard graduates from *Chongryun* high schools as being eligible to sit for the entrance examinations to state universities. This discriminative act was apparently spurred by the problems in Japan's relations with North Korea, such as the admission by Pyongyang of having kidnapped Japanese citizens.

Another ordinance that discriminated between the pro-North Korean and the pro-South Korean diaspora group was enacted in August 2006, when the Japanese Ministry of Justice announced a restrictive policy on non-South Korean nationality holders, reducing the duration of their reentry permits from four years to a maximum of one year. It also made it mandatory for them to report the purpose and itinerary of their trips when applying for re-entry permits.[140] The diaspora card that was useful for expanding Japan's autonomy was no longer a necessary tool. Closing the gate for the pro-North Korean diaspora to cross boundaries became more important for the Japanese Ministry of Justice. Despite having identical permanent residence, South Korean nationality holders were not sanctioned in this way.[141]

In order to curb pro-DPRK activities in Japan, Japanese lawmakers introduced legislation to restrict, and ultimately ban port visits by DPRK ships such as the *Mangyongbong-92*, which were suspected of illicit activities. In late January 2003, Kim Sang Gyu, a former *Chongryun* senior official, confessed to being a North Korean agent and running a spy network in Japan until 2000 which included instructing collaborators to gather political and military data on Seoul.[142] The police also found in his home important documents, such as order letters to recruit sympathizers in Japan (and South Korea), with specific datelines for the recruitment period, as well as drafts of letters addressed to North Korean officials. Kim confessed that he received espionage orders from Pyongyang via the *Mangyongbong-92*.[143] Furthermore, a former North Korean missile scientist testified on May 20 2003 at a US Senate hearing that more than 90 per cent of the components used in Pyongyang's missile programme were smuggled in from Japan by *Chongryun* aboard the *Mangyongbong-92*. He stated 'the way they bring this in is through ... the North Korean association inside Japan; they bring it by ship every three months.'[144] The ferry runs between Niigata and Wonsan on a bi-weekly basis and serves as an important vehicle for re-uniting repatriated family members in North Korea, for official exchange of gifts, letters or financial assistance, and for political participation of elite association members in North Korea's politics. The 1,000-ton cargo on this vessel (and the 200 passengers) had not in the past undergone the strictest scrutiny and had legitimately transported export goods and humanitarian aid supplies based on Japan's laws.[145] The ship transported Zainichi Koreans who were visiting their relatives, *Chongryun* school children who were on study trips, and necessary goods such as electronic devices, medical instruments and other products to North Korea.[146] In 5 July 2006, the Japanese government finally banned the *Mangyongbong-92* from entering any Japanese port.

In September 2003, Tokyo Governor Shintaro Ishihara decided to end *Chongryun*'s tax-exempt status as rumours surfaced alleging links to smuggling and illegal exports of missile parts.[147] Ironically, prior to that, the liberal Tokyo Metropolitan Governor Minobe (who held the same rank as Shintaro Ishihara) decided not to ask the pro-Pyongyang group to pay fixed-asset and urban-planning taxes, saying it recognized that its properties were being used for a state delegation in Japan, and as such were exempt from taxation under the Vienna Convention.[148] Only 10 months after North Korean leader Kim Jong Il admitted in September 2002 that the country had kidnapped 13 Japanese in the 1970s and 1980s, Tokyo Governor Shintaro Ishihara had about 42 million yen in taxes imposed on the group for the 2003 tax year. In December 2003, the Tokyo Metropolitan government filed a lawsuit demanding the return of a 4,140-square-meter portion of the 5,400-square-meter plot that *Chongryun* occupied and had been allowed to use free of charge until 1990.[149] In the suit filed in May 2004, *Chongryun* representatives said the properties had never been taxed and Ishihara's decision to change this policy was 'discriminatory treatment that has no legitimate reason.' But the court rejected that claim, stating that the metropolitan government had the right to change its decision and had exercised that right in an appropriate manner.[150] Other municipalities where the *Chongryun* holds properties – it has at least 300 local chapters nationwide – have made a range of decisions on how to treat the group. It has been estimated that the *Chongryun* has 50,000 North Korea-affiliated members nationwide, and its Tokyo headquarters has been in the spotlight since the Tokyo District Court in June ordered the pro-Pyongyang group to pay 62.7 billion yen in debts to the government-backed Resolution and Collection Corporation. The verdict permitted the headquarters building and property to be seized and the process has begun for the RCC to sell it.

Japanese public opinion and media effectively served in helping Japan's exclusionary policies towards the pro-North Korean diaspora. The revelations based on testimonials from a former *Chongryun* official and from a defected North Korean scientist, which were widely reported in the Japanese media, directly linked North Korean espionage and clandestine activities with *Chongryun* in the minds of the Japanese public, regardless of the facts behind the allegations. Public sentiments turned against *Chongryun* due to its connection with North Korea. Certain government officials took advantage of this situation and not only tried to undermine *Chongryun* but also promote their own extreme revisionist foreign policies in line with Japan's right-wing politicians who were rejecting postwar pacifism, embracing the old imperial system, defending Japan's past wars in Asia and even promoting the idea of a nuclear Japan.[151] Although *Chongryun* served as an unofficial embassy for North Korea, its headquarters and affiliated regional offices have received numerous threats and become the targets of demonstrations by Japan's right-wing and anti-communist groups.

Some civil rights movements used to protest in favour of the Zainichi diaspora's inclusion into Japanese society, but now other kinds of civil movements have emerged with full-fledged public support, which favour certain nationalist groups that are involved in other right-wing causes, such as Vice Chairman Yoichi Shimada, the Japanese Society for History Textbook Reform and former Chairman Katsumi Sato, who promotes the idea of a nuclear Japan.[152] The National Association for the Rescue of Japanese Kidnapped by North Korea (NARKN or Sukukai hereafter) was established not only to support the kidnapped Japanese but also to protest against North Korea and *Chongryun*. In early 2005, Sukukai filed a lawsuit against Kumamoto public officials for granting tax benefits to *Chongryun* for running its Kumamoto Korean Hall. Although the Kumamoto District Court ruled in *Chongryun*'s favour, the group appealed to the Fukuoka High Court, which decided in February 2006 that *Chongryun*'s work does not benefit the general public.[153] The presiding judge, Hiroyuki Nakayama, stated that *Chongryun* 'conducts activities to benefit Korean residents of Japan under the leadership of North Korea and in unity with North Korea and is not an organization that in general benefits the society of our country.'[154] The Japanese government allowed *Chongryun* to operate relatively freely until the late 1990s despite the fact that since 2002 the recognition of North Korea as a threat has been spreading throughout Japanese public consciousness and has given nationalists and right-wing group new grounds for demolishing *Chongryun* under the pretext of protecting Japan's national security. *Chongryun*'s presence provides the Japanese public, especially nationalists, with a link to the North Korean government and creates a kind of xenophobic emotional perception by bringing Japan's enemy or national security threat onto its home shores.[155]

Finally, pro-North Korean diaspora members have gradually come to the realization that North Korea is not the beloved homeland, and have been disappointed at the North Korean leadership's failure to display any concern whatever toward those Zainichi Koreans who have supported the regime financially and morally for over half a century. Such a realization among pro-North Korean communities allowed them to reveal secret family stories that were not usually told publicly before. *Our Homeland* (Kazoku no Kuni) is a story of one Zainichi diaspora member's own life, which was screened in the Forum section of the Berlin Film Festival, winning the CICAE Prize in 2012. Yang Yong Hi was born and raised in Osaka's *zainichi* (ethnic Korean) community, and debuted as a director with 'Dear Pyongyang,' a 2005 documentary telling the story of her father, who emigrated to Japan from the southern Korean island of Cheju, but after the partitioning of the country in 1945 became a fervent supporter of North Korea. *Our Homeland* tells the story of one of her brothers, who returns for just a short period. Although Ms Yang and her parents were allowed to visit North Korea occasionally, her brothers were prohibited from leaving the communist state, with the exception of her youngest brother's medical visit. The film of *Our Homeland* implies that the homeland of this family is in the final analysis neither in the North nor the South, but in a past that can never be brought back, set against a future that promises more of the same.[156] When Ms Yang spoke with the Journal and was asked about how she considers her homeland, she answered that 'Our Homeland,' refers to a certain place, not a nation; that is the ideal place where a family can truly be together.[157] Ms Yang's siblings were among the more than 93,000 ethnic Koreans estimated to have been shipped to North Korea from Japan between late 1950s and early 1970s. Ms Yang's conclusions may be shared by these repatriated Zainichi families. A member of one of the other repatriated families to North Korea, whose identity as a Zainichi Korean was hidden in her daily life, also revealed tearfully:

> When my family repatriated to North Korea, we already knew the reality of North Korea, but we cared about our own family and their safety too much. If we, all Zainichi Koreans, honestly talked at that time, we would not have suffered for so long until now. Please tell me how I can meet my brother again. I want to meet him freely.[158]

Conclusion

During the 1970s and 1980s, despite strong South Korean protests to crack down on the pro-DPRK groups in Japan, Tokyo refused to do so unless these groups could be proved to have engaged in illegal behaviour. Even in 1974, when Mun Se Kwang's assassination attempt on Park Chung Hee killed the South Korean president's wife, the government of Japan refused to launch a pervasive crackdown against all *Chongryun* activities in violation of their civil liberties. Furthermore, the end of the Cold War and the 1993–4 North Korean nuclear crisis led the US ask for Japanese help in preventing pro-DPRK residents from sending remittances back to North

Korea. The Japanese reaction was not as rapid or all-encompassing as Washington expected, and the National Police and Finance Ministry was warned not to stop money flows to Pyongyang.

So how do we explain such a change in approach by the Japanese government towards the pro-North Korea diaspora? What conditions changed to allow the host state to treat the diaspora group in a different way to that which the hypothesis suggests? As Morrow states, the clue to explaining the difference relates to whether the relations among allied members are asymmetric or symmetric. From the Cold War years until the earlier stages of the post-Cold War era (1990–2002), the model of alliance was asymmetric. Indeed, both the Japan–US security alliance and the US–ROK alliance were heavily dependent on the US market and American security guarantees during the Cold War era; the pattern of these alliances can definitely be described as asymmetric. If we compare US power during the Cold War or from 1990–2002 with the period from the autumn of 2002 until the present, we cannot ignore the signs of a relative decline in US hegemony in comparison to the rise of China; as such it is necessary to check the degree to which the US can still play the role of major state in the context of the alliance framework.

As a minor state, the ROK had, by the mid-1990s, already qualified for membership into the Organisation for Economic Co-operation and Development (OECD), formally joining the world's roster of affluent nations and including itself among a privileged group of highly industrialized constitutional democracies.[159] The country now plays an influential role as a middle power or 'balancer' in East Asia. South Korea is no longer afraid of North Korea's unilateral military threat, but would still tread very carefully towards preparing for re-unification with the North. South Korea cannot neglect the growing importance of China for its further economic prosperity or its potential unification with North Korea. China – North Korea's biggest trading ally – is the nation believed to wield the most influence over the government in Pyongyang, yet in July 2014 Chinese President Xi Jinping's decision to visit Seoul before Pyongyang is being seen as a typical sign of the close relationship between South Korea and China, as Chinese leaders have traditionally visited Pyongyang before Seoul.[160] South Korea and China also reaffirmed their opposition to any further nuclear tests by Pyongyang, and Xi and Park also agreed to conclude negotiations for a free trade deal by the end of the year and take measures to spur offshore use of the yuan and investment in the Chinese markets. The two countries already have a strong commercial relationship with two-way trade worth an estimated $230 billion (210 billion euros) annually. 'China and South Korea as neighbors must jointly respond to challenges in the security environment, while sharing the opportunity for development that peace and stability of the region offers,' said Xi in an article that was carried by all major South Korean newspapers.[161] Hwang Byung-tae, a former South Korean ambassador to China, believes that the two leaders might have very different ideas about what is important right now since China's interest is moving away from the North Korean issue anyway. 'South Korea is strategically important [to Beijing] because of China's relationships on its eastern boundary,' he said. 'First there was America's pivot to Asia, then Japan's new nationalism, there's old rancor with Vietnam and territorial issues with the Philippines. South Korea lies in an important position and friendly relations with it are part of China's grand scheme.'[162]

As another minor state in terms of its domestic legal and military restrictions, Japan is now trying to pave the way for changing the traditional interpretation of Article 9 to enable Japan to exercise its right to collective self-defence by calling for 'proactive pacifism,' which would enable the country to stand as a normal ally with the US. By citing external threats, the Abe administration now seeks to consolidate domestic support in this direction. Most recently, on 1 July 2014, the Japanese Prime Minister's administration announced a major new interpretation of the security provisions of the country's 1947 constitution, permitting its SDF to participate for the first time in collective self-defence related activities.[163] Japan has approved reinterpreting the country's pacifist Constitution to allow the use of the right to collective self-defence in a major overhaul of postwar security policy. Although Japan's defence forces have been unable to extend their military collaboration with their US allies beyond a narrowly circumscribed role, in future the SDF will in principle be able to assist the forces of a foreign country in situations where either the survival and security of Japan or that of its citizens is at risk. Examples include providing defensive support to US forces under attack in the vicinity of Japan, co-operating militarily with US forces to safeguard Japanese citizens at risk overseas, participating in minesweeping activities during times of war and deploying Japanese forces to protect access to energy supplies or critically important sea-lanes vital to Japan's survival. This will provide Japan with much greater latitude to strengthen its military ties with the United States. It will also potentially open the door to more active defence co-operation with other countries in the Asia–Pacific region, such as Australia and the Philippines – both of which have welcomed these changes as they look anxiously at China's increasingly assertive maritime posture in the South and East China Seas.[164]

Such new measures look set to further undermine Japan's already frayed relationship with South Korea and to heighten territorial and political tensions with China. Indeed, the security environment surrounding Japan seems very threatening when viewed from a Japanese perspective. China continues to challenge the international maritime order. With the decline in US influence, there are uncertainties about how to proceed with the Senkaku Island's dispute and problems concerning North Korea's nuclear and missile development programmes. US Defense Secretary Chuck Hagel praised Japan's efforts, saying they would help 'build a peaceful and resilient regional order' at the Shangri-La Dialogue international security forum in 2014. Generally, the new interpretation is likely to strengthen the perception that Japan has become a comparatively normalized state in terms of its ability to contribute constructively to global and regional security. Thus the asymmetric trilateral alliance between the US, Japan and South Korea has gradually shifted into a closer and more symmetric alliance in the later post-Cold War era.

Subsequently, because of the structural transition of the alliance from asymmetric to symmetric, Japan's policy towards the pro-North Korean diaspora exactly matches Mylonas' nation building theory. In response to the external threat from an enemy country, North Korea, Japan has employed a stricter exclusionary policy to that part of the Zainichi diaspora group which is supported by North Korea. Only under an asymmetric alliance will alliance cohesion matter in terms of policy making on the part of the host state towards a non-core diaspora group in terms of exclusion, assimilation or accommodation. Minor powers have low levels of security and high levels of autonomy and will therefore try to form alliances that increase their security at the cost of some autonomy. The evaluation of the trade-off between autonomy and security for a particular nation depends on how it separates the various issues under those specific headings.[165] As a substitute for the real nation state, a 'de-territorialized entity' or a diaspora is likely to compromise the inequality between major states and minor states in terms of rational calculation regarding the trade-off between autonomy and security, which will allow the host state to manoeuvre without breaking alliances in order to expand autonomy or gain security assurances.

Under symmetric alliance conditions, however, the situation is very different. Each allied member possesses high levels of both autonomy and security. Members have no overriding interests to enhance either of these factors. Alliances last only as long as they are useful against the threats they encounter. But alliances find it much more difficult to share harmonious and divergent interests when the capabilities of allied nations change. Alliances will be terminated when the threat passes. Under symmetric alliance, every allied member is likely to react directly and conduct policies toward non-core diaspora groups in response to an external threat. It is especially hard to keep what Victor Cha terms a 'quasi-alliance' going in the face of the geopolitical uncertainties that currently pertain in East Asia.

Indeed, as Mylonas explores, the way a state treats an assimilated ethnic group within its own territorial boundaries is determined largely by whether the state's foreign policy is revisionist or complies with the international status quo, and whether it is allied with or in competition with that group's external patron. Mylonas quite rightly pays particular attention to external involvement in cultural differences and plans for nation building rather than domestic national factors in terms of cultural difference. However, by focusing on state capacity and its power configuration rather than discussing bilateral relations within alliances, the asymmetric alliance in the East Asian region that emerged at World War II shows that such asymmetric alliance bloc structure sometimes allows the less powerful host state to accept external involvement in reflection to its own foreign policy. It has, however, been conducted with the permission of the security umbrella held in place by superpower allied state.

After the end of World War II and the subsequent Cold War, the role of the United States as a major power was indispensable for the stability of Japan's postwar survival with its constitutional policy of restraint on the use of military force, and for the quick recovery for the devastated South

Korea after the Korean War under a state of armistice on the Korean peninsula. Such asymmetric power relations allowed Japan to focus mainly on the Japan–US relationship in the same way that they allowed South Korea to focus on the US–ROK relationship. This explains why alliance cohesion matters to external involvement on cultural differences and on the politics of nation building in the context of the US–Japan and US–ROK asymmetric hierarchic alliances. For Japan, during the Cold War era, in the years when US sometimes demanded tightness in terms of security alliance, cultural blindness allowed Japan to recognize China or North Korea as a Communist group rather than as ethnic Chinese or ethnic Koreans, and employed strict exclusionary policies toward the pro-North Korean diaspora. But after the end of the Cold War, the Japan–US relationship was no longer the only relationship that mattered. Japan was increasingly forced to deal with Asia and China and their demands, which compelled the Japanese government to wake up from cultural blindness – or what Carol Gluck called 'geopolitical blindness' – and forced Japan to adopt a more revisionist foreign policy.

However, recent young and unmarried Koreans have increasingly begun to align themselves with a new identity that is not based on a particular ethnic culture. In other words, they do not attempt to identify themselves as belonging to South or North Korea or Japan, although they still retain their Zainichi identity, which literally means 'staying in Japan'. They prefer to call themselves simply Zainichi, rather than Zainichi Korean. In the end, the Zainichi diaspora becomes 'the nationalist genie, never perfectly contained in the bottle of the territorial state' as Appadurai claims.[166] This is the consequence of a geopolitical power game; external involvement by great powers in the process of nation-building. But Homi Bhabha argues that 'hybridity' is the third space from which other positions may emerge. The third space displaces the histories that constitute it and establishes new structures of authority, new political initiatives.[167] Rather than falling into the bipolar identity classification of being Korean/Japanese or South/North Korean, the third way for the Zainichi identity might provide new possibilities for the reconciliation. Kang explains that there are two schools of thought for some of the younger generation Zainichi. One is that the country they love most is Japan and the one they most dislike is the Korean peninsula. The other is that the country they love most is the Korean peninsula and that which they dislike most is Japan. Kang continues that this sense of division in the Zainichi identity must be healed through reconciliation. Kang concludes that the process of reconciliation – the unification and co-existence of North and South – is not simply a matter of bringing together a divided country, but will also be bound up with simultaneous reconciliation with Japan.[168]

Conclusion

Geopolitics and Shifting Policies towards the Zainichi Diaspora

Most overseas Koreans are concentrated within five countries: the United States (2 million), China (1.9 million), Japan (0.6 million) and Commonwealth of Independent States or CIS (0.5 million).[1] These countries account for more than 90 per cent of all overseas Koreans, and most Korean emigration is historically related to Japanese imperial expansion, with the exception of the US. Among Korean émigrés, the Zainichi diaspora is of particular interest because they are the only migrants who have not been granted citizenship among co-ethnics abroad, such as ethnic Koreans in China and in Russia, as well as Korean Americans. Although they have become highly acculturated to Japanese society over time, there are still around 400,000 Zainichi who maintain their Korean nationality in Japan, although they do not intend to repatriate to Korea. What forces govern the contradictory identity of the Zainichi, a group with legal claims to a homeland yet who remain culturally Japanese?

Existing literature on Zainichi Koreans focuses mainly on ethnic minority aspects within the context of the history of Korean minorities in Japan, and discusses ethnic discrimination and comparison with other minority groups or the formation of an ethnic identity at national level.[2] This book argues that policy making towards specific ethnic groups such as the Zainichi Koreans is not driven simply by historical antipathy or by the significance of inequalities in resource distribution, social benefits and opportunities between a particular ethnic group and the dominant national group, as most scholars argue. What is conspicuous by its absence from the literature is an analysis of the ways in which external factors such as the changing influence of a diaspora's homeland can affect minority behavior or the attitudes of the states in which those minorities live. Moving from a domestic focus of 'ethnic' or 'non-citizen' groups to a transnational focus when studying a diaspora allows for new perspectives that can incorporate geopolitical contexts. Diasporas are among the most prominent actors linking international and domestic spheres of politics. Because diasporas possess both external and internal contexts – insofar as they have a home country and a host country – they can become a political instrument for both sides. The particular angle I identified in this book explores how geopolitical developments impact on a diaspora. From the standpoint of international relations, the book accounts for the ways in which diasporic configurations such as the Zainichi become enmeshed in geopolitical relations and how those relations have affected the shifts in Zainichi identity. In terms of geopolitics, I examined the nature of the alliance system that persisted for a long period after World War II and played a key role in influencing the region's international relations in East Asia.

Interstate relations between host state and external homeland can explain why among overseas Koreans only the Zainichi diaspora is split by differing opinions on their divided homeland. Mylonas has shown that interstate relations between the external homeland and host state are important in terms of the host state's policies governing the ethnic group, depending on whether the homeland with which the ethnic group identify their essential belonging is part of the same alliance bloc or not.[3] Mylonas' nation-building theory analysis is applicable when the North and South Korean states clearly act as a homeland – such as with North Korea's wholehearted commitment to mass repatriation of Koreans in Japan in 1958 and the normalization treaty between Japan and South Korea in 1965. It is also relevant when Japan perceives the triangular asymmetric alliance among US, Japan and South Korea as vital to its own national interest and security.

What conditions govern the inclusivity or exclusivity with which states treat particular ethnic groups? Building on the logic of Mylonas' nation-building theory, I focussed more specifically on the cohesion of alliance. The other indispensable factor emphasized in this work is what Brubaker refers to as a 'category of practice' regarding the diaspora. Brubaker identified a diaspora as a useful group with which the host state can practise the use of flexible tools for increasing its autonomy.[4] From examining the literature and studying existing theories, I generated three theoretical hypotheses for the possible pathologies of nation-building policies:

Configuration 1: if alliance bloc cohesion is strong, the host state policy will be one of *exclusion* toward a diaspora group supported by an enemy state, but one of *inclusion* toward a diaspora group supported by an allied homeland.

Configuration 2: if alliance bloc cohesion is weak, the host state (Japan) will *accommodate* a diaspora group supported by a homeland belonging to an enemy alliance bloc in order to expand its own autonomy.

Configuration 3: if the host state and homeland have no diplomatic relations, then the host state policy toward the diaspora group perceived to be backed an enemy homeland will be *exclusion*.

The Cold War era and the post-Cold War period provided excellent scope for testing my hypothesis by using the process tracing method. Alliance cohesion is tested when the critical exogenous geopolitical situation shifts, at which point the home state will consider 'Who will support whom and who will resist whom, and to what extent?' At this point the space occupied by the diaspora moves, either by expanding or contracting, as a reflection of the policies employed by the host state: exclusion, assimilation or accommodation.

Factors Identified by the Zainichi Diaspora Case Study

Three important factors result from this hypothesis testing. Firstly, during the Cold War era alliance cohesion was important in terms of policy strategy towards diaspora groups by host states. During the Cold War, the measurement of alliance cohesion was the independent variable that resulted from policy strategy by host states towards diaspora groups. Under the strong US security umbrella, occasional shifts in alliance cohesion lent themselves to generating space occupied by the Zainichi, either by expanding or contracting its power and influence in relation to the policies employed by the host state; exclusion, assimilation or accommodation. In the post-Cold War years, however, the North Korean threat became a more significant factor, so the degree of threat perceived by the host state had a strong impact on policy making toward the diaspora. Ultimately, we found a structural change of alliance from asymmetric to symmetric.

Secondly, the accommodation policy toward an enemy allied diaspora by a host state is driven principally by state interest. A possible alternative explanation for ambivalent attitudes towards an enemy allied diaspora group by Japanese elites is that Japan's political leaders sought to avoid confrontation with *Chongryun* (the pro-North Korean diaspora association) out of fear of provoking reaction from Pyongyang and out of sensitivity to accusations of racial discrimination, especially from international society. As I explained in Chapter 4, however, my own belief is that the Japanese government's ambivalence towards the North Korean regime was driven by Japan's two-Korea policy. Rather than highlighting this sensitive policy, more relevance was officially accorded to the policy of *Seikei Bunri* (separation of economics from politics), under which Japan officially recognizes the ROK but not the DPRK, while continuing to trade with both. The two-Korea policy went hand-in-hand with the *Seikei Bunri* approach, and inevitably affected Japan's treatment of Korean nationals. With no official normalization between Japan and North Korea, the *Chongryun* headquarters was used as a *de facto* embassy for North Korea, allowing Japan to promote an accommodation policy by maintaining the cultural specificities of the diaspora group. During the Cold War era, North Korean factors became a useful political instrument, a fact that underlines one of the main arguments of this thesis, namely that weak alliance cohesion allows the host country to accommodate a diaspora supported by an enemy alliance group in order to expand its own autonomy under an asymmetric alliance (**Configuration 2**).

The third important factor is that in order to employ an effective accommodation policy, it is vital to build up the mechanism of interstate alliance at the same time. The accommodation policy employed towards the

CONCLUSION

Zainichi diaspora group supported by an allied homeland (South Korea) was more effective in reducing the inclination of diaspora mobilization towards their own homeland than the frequent accommodation policies employed in relations with the pro-Pyongyang diaspora group during the Cold War. Chapter 4 introduced my argument surrounding **Configuration 1**: if alliance cohesion is strong, the host state policy will be one of *exclusion* to a diaspora group supported by an enemy allied homeland, but one of *inclusion* to a diaspora group supported by a common allied homeland. After fourteen years of extremely difficult negotiations, the governments of South Korea and Japan signed the seven-article Treaty on Basic Relations on 22 June 1965, as well as other agreements, protocols, notes on property claims and economic cooperation, fisheries and cultural assets. In terms of cost benefits from the bargaining calculations, the Japan–South Korea diplomatic normalization years (1965–70) provide evidence that the strong cohesion of the US–Japan–ROK alliance produced a harmony of interest between all parties by maintaining a balance between security and autonomy. But at the same time the theory supports the argument that host state policies will only accommodate an allied diaspora group under conditions of strong alliance cohesion. Only those who registered as allied nationals received permanent resident status in Japan, and became entitled to rights and benefits that had previously been closed to them. With this status they would be guaranteed protection under Japanese law, they would be exempt from deportation and would be able to join Japan's national health insurance scheme.

At the outset, however, the treaty was not welcomed by many Zainichi Koreans in Japan. Citizenship of South Korea was considered little more than symbolic – a clear verification of the bearer's alignment with South Korea over North Korea. The policy was seen as being aimed at producing as many Japanese Koreans as possible who also possessed a South Korean passport, rather than welcoming the diaspora as a whole into Korean co-ethnicity. Having said that, after the treaty was signed, many Koreans began officially changing their citizenship and, by 1970, 52 per cent of all Zainichi Koreans in Japan had registered themselves as South Korean nationals, a figure that is growing to this day.[5] This was in sharp contrast to those who did not register as South Korean nationals – the pro-North Korean diaspora. The exclusionary policy set up a strong link between the Zainichi diaspora and North Korea, yet it did little more than keep them in a kind of legal limbo, at least until 1982, when Japan created a new category of permanent residence for them, called Exceptional Permanent Residence. This meant that Japan had allowed the North Korean diaspora to exist in an ambiguous legal position for sixteen years.

For several decades after World War II, East Asian regional state architecture was characterized only by alliances that could promote de-politicized ethnic divisions between two antagonistic countries such as Japan and South Korea, by creating incentives for everyone to support mutual regional security against an enemy communist bloc. In fact, it took until 1965 for a bilateral treaty between Japan and South Korea to be agreed due to the deep-seating antipathy of the Korean people as well as President Rhee's inability to adjust to the reality that Japan was no longer an enemy. The US played a significant role in constructing a regional interstate alliance between colonizing and colonized countries, as it was essential that Japan overcome the historical enmity between the two countries for its own security as well as that of the US–Japan alliance. According to archival documents, it is clearly recorded that 'turning to Korea, the Under Secretary said that he had hoped that normal relations with Korea [would resume].There is much, he said, that Japan can do which the US cannot because of Japan's business ties with and knowledge of Korea and Japan's experience in agriculture.'[6]

Accommodation Policies towards Divided Diaspora Groups and Variation in Outcomes

Japan's integration policy from 1965 onwards gradually brought more tangible effects to the relationship between Japanese society and the Zainichi diaspora, allowing the Zainichi to gravitate toward South Korea, a move that has minimized its potential for mobilization towards its own homeland. Firstly, although the South Korean government began funding educational programmes at *Mindan* schools in 1957 to counter propaganda coming from the North Korean regime to Korean communities in Japan, the majority of Zainichi diaspora who possess South Korean nationality still go to Japanese schools. Unlike *Chongryun*, however, most *Mindan* schools are not for the children of Zainichi Koreans, but for the children of South Korean diplomats and businessmen who are temporarily living in Japan.

Secondly, in comparison with the *Chongryun*, the scale of *Mindan*'s monetary transfer is insignificant. There have been exchanges of gifts and financial assistance between *Mindan* and the South Korean government. For example, *Mindan* sent a total of $52 million to South Korea between 1963 and 1995, aimed at supporting the 1988 Seoul Olympics ($45 million), various construction projects ($2.5 million) and as relief for national disasters in South Korea. Yet the scale was far smaller than the flow of money, technology and know-how that flowed between *Chongryun* and the North Korean regime. *Chongryun* was at the heart of a fund-raising effort that sent at least $100 million to North Korea every year.[7] Furthermore, the *Chongryun* donated $32 million to the construction of a 9,672-ton cargo-passenger ship, the *Mangyonbong,* for Kim Il Sung's eightieth birthday in 1992. Unlike the *Chongryun*, *Mindan* do not rely on repatriated families, as members can freely visit and leave South Korea.[8]

The accommodation policy employed towards the pro-North Korean diaspora group was motivated mainly by Japan's own national interests rather than a desire to prioritize regional stability, and came about because of a temporary weak alliance cohesion. However, when facing external threats, all allies inevitably abandon nationalist policies that seek to achieve individual countries' desired ends, and are instead required to prioritize the elimination of conflicts of interest that could separate the allies. These shared goals – such as regional peace and stability – would contribute to strengthening and consolidating the cohesion of the alliance between the US, Japan and South Korea. Under conditions in which there is no interstate alliance it becomes more difficult for governments to maintain a policy of accommodation towards a diaspora group supported by an enemy state, because from a government perspective, regional stability is likely to be more important than national interest.

To this day, *Chongryun* has officially functioned as an association for overseas North Korean nationals. The North Korean nationality ascribed to *Chongryun* Koreans is in fact an entirely manufactured and symbolic identity. Although Pyongyang declared Koreans in Japan citizens of the DPRK, no consular services ever existed and no passports have ever been issued. *Chongryun* Koreans have never voted in North Korean elections, nor have they been conscripted into the North Korean army.[9] However, the term 'overseas nationals of North Korea,' a term central to the *Chongryun*'s legitimate discourse, became an identity that effectively replaced that of colonial subject. Indeed, the alliance cohesion has fluctuated over time in response to shifting regional and geopolitical power configurations, with the result that the policy towards the pro-North Korean diaspora has oscillated between accommodation and exclusion. The policy makers subsequently lost the opportunity to integrate this unassimilated non-core ethnic diaspora in a way that would minimize the capacity of the diaspora to mobilize in support of their own homeland.

Rather than moving to a path of integration into Japanese society under the national integration guidance, *Chongryun* Koreans were placed in a position of self-contained co-existence within the Japanese state, in which they kept their dialogue with Japanese authorities to the absolute required minimum.[10] The frequent accommodation policies employed towards them by Japan during the Cold War era – especially when the US commitment to the Asia–Pacific region became weak – had in fact been driven by this bargaining deal between Japan and *Chongryun* under the guidance of North Korean leader Kim Il Sung since 1955. Indeed, as of early 1955, *Chongryun* leadership was successful in solidifying its members as North Korean nationals based on the rhetoric that the Great leader Kim Il Sung's wise guidance had allowed Zainichi Koreans to become dignified overseas nationals of their fatherland and prevented them suffering all sorts of oppression by 'enemies' – including US imperialism, the South Korean puppet regime that North Korea considered corrupt and oppressive and the Japanese authorities who were assisting them. By emphasizing the fact that *Chongryun* activities are practiced and organized in the service of Kim Il Sung and the fatherland, *Chongryun* officials could effectively appeal to their affiliates' patriotism while self-righteously denying any interest in personal advancement, claiming instead to work for the sake of the fatherland, the nation, and compatriots at large.[11] Ultimately, the *Chongryun* began to represent itself as North Korea's 'money pipeline,' especially after the collapse of the Soviet Union and the cutbacks in Chinese aid to North Korea.[12] This 'money pipeline' has become the subject of intense debate among US officials, especially concerning the question of how much money is involved and where it goes (to individuals or to the regime).

Under the rigid political ties between *Chongryun* and the North Korean government, combined with Japan's accommodation policy, the system of hard currency flowing into North Korea was established and operated continually until the test launch of the Taepodong missile over Japan in August 1998 by North Korea and the admission by Kim Jong Il in September 2002 that North Korea had abducted several Japanese nationals. Ultimately, Japan's defence posture has been significantly affected by the rise of the perceived North Korean threat since the mid-1990s, provoking a shift from low-profile postwar security policies towards a position as a more normal military actor and US alliance partner.[13]

Policy Implications

What lessons can we learn from the examples provided by the Zainichi diaspora? Recent studies explore how the perceived threat from North Korea helped compel Japan to enhance its defence posture,[14] yet very few of those studies have shown how much of the material resources for the North Korean military regime was raised among the diaspora population within Japan. These funds and resources became significant factors in contributing to the North's missile and nuclear capabilities, which in turn exacerbated the political and military threat to the solidarity of the US–Japan alliance and pact, threatening to undermine the very foundation of Japan's postwar security policy. In other words, the way in which the North Korean threat is intimately related to Japan's policy towards the divided diaspora populations remains underexplored.

Among various policies employed by Japanese governments, the exclusionary policy towards the Zainichi diaspora – more specifically its repatriation programme – formed the starting point for a diaspora mobilization movement and thus became the basis for the kin of the Zainichi diaspora to create a tangible link with North Korea. As mentioned earlier in this chapter, among overseas Koreans only the Zainichi diaspora is split by differing opinions on its divided homeland, a fact that is related to Japan's exclusionary policy and exemplified by the repatriation programme. From the perspective of the Japanese authorities, there was no great difference between the two Koreas, as Japan had no formal diplomatic ties to either of the countries. This also stands in support of **Configuration 3** in my argument: if there is no alliance between host state and homeland, the host state policy towards diaspora groups from an enemy allied or a common allied homeland is *exclusion*.

In fact, the Japanese government supported the repatriation programme as a way of reducing its own economic stress as well as to rid the country of ethnic minority residents who were regarded as indigent and vaguely communist.[15] Furthermore, the Japanese government also required Koreans who were repatriated to North Korea to renounce their right to re-entry into Japan when they left. This exclusionary policy, however, changed in 1965 when Japan and the Republic of Korea normalized relations, and that was the decisive moment when the divided diasporas were generated within the host country, Japan. Yet those 93,000 Koreans who were repatriated to North Korea between 1959 and 1984 wound up becoming instruments by which diaspora entrepreneurs such as *Chongryun* deployed framing systems emphasizing guilt or obligation, such as sending remittances to family members.

As previously mentioned, a significant percentage of the funding for material resources for the North Korean regime came from the diaspora population within Japan. Until late 2002, Japan had been a significant source of North Korea's foreign exchange and an important hub for transferring high-tech equipment to North Korea that could be used to develop the North's nuclear weapons programme. *Chongryun* has been implicated in these transfers as a North Korean agent in various ways.

Firstly, remittances from Japan to the DPRK are the single most important source of hard currency. Before Japan's economic bubble burst in the early 1990s, US officials estimated that up to $1 billion a year could have been transferred to North Korea from Japan, but by all accounts, that figure has decreased with the economic downturn in Japan and the death of Kim Il Sung.[16] Most observers now agree with Nicolas Eberstadt's estimate that even at their peak, remittances were generally below $100 million annually.[17] In terms of the main resources for the remittances, Apichai Shipper argues that the North Korean regime has been effectively utilizing Koreans who were repatriated to North Korea as an instrument to maximize loyalty from family members who remain in Japan as members of *Chongryun*.[18] Former residents of Japan who moved to North Korea between 1959 and 1985 under relocation campaigns regularly receive money and goods from their relatives in Japan as aid packages.[19] In addition, if *Chongryun* Koreans wanted to invite their repatriated families over to Japan, they had to pay approximately 2 million yen (US $20,000 at the exchange rate of $1:100 yen) for a one-week visit.[20]

Illegal remittances to North Korea through several credit associations such as the *Chogin* (Korean bank) also contributed heavily to financial support for the DPRK. The banks were accused of allowing Pyongyang sympathizers to use fictitious or borrowed names to create bogus accounts, which were then used to channel cash to North Korea, as well as offering preferential loans to people who donated large amounts to the North Korean cause.[21] The banks also lent money in excess of the collateral, which was often land and buildings either owned by or affiliated with *Chongryun*.[22] In the end, most of Japan's *Chogin* credit unions collapsed when the banks began failing, as their nonperforming loans increased, necessitating an infusion of public funds to protect ordinary depositors. Japan's Financial Services Agency authorized the transfer of the operations of several failed credit unions to four new lenders with the stipulations that the unions sever ties with *Chongryun*. More than $3.3 billion in public funds was pumped into the replacement credit unions in 2002 alone.[23]

Secondly, the Japanese market was a major destination for the North Korean government's suspected drug-running operations. President Bush, in his annual report to Congress, asserted that state agents and enterprises in the North were involved in the narcotics trade. The *Mangyongbong* passenger ferry was suspected of carrying illicit shipments of drugs, although the boat was also used to transport Zainichi Koreans who were visiting their relatives to North Korea. Such international drug-trafficking between Japan and North Korea may have provided up to $7 billion in cash profit for the North Korean regime.[24] According to the *Yomiuri* newspaper, a North Korean escapee said that he had smuggled narcotics for the regime on the ferry, handing over the goods to couriers who were Korean residents in Japan, who would then pass the drugs on to Japanese gangsters for sale.[25] Reported indications show that ties between North Korea and the Japanese gangsters are on the rise, although documentation of links specifically with *Chongryun* remains scant.

Thirdly, *Chongryun* has also been involved with at least one other non-ethnic Korean company selling military technology to North Korea. Some

Japanese firms associated with the *Chongryun* have been implicated in illegal plans to transfer high-tech equipment to North Korea that could be used to develop the North's nuclear weapons programme. One example is Meishin, a Tokyo-based trading company run by *Chongryun* members which has admitted that it exported three transformers to North Korea via Thailand, instruments that regulate electrical current and can be used for uranium enrichment as well as missile development.[26] Also, a former North Korean missile scientist testified on 20 May 2003 at a US Senate hearing that more than 90 per cent of the components used in Pyongyang's missile programme were smuggled in from Japan by *Chongryun* personnel abroad the *Mangyongbong*.[27] A senior member of a science and technology organization affiliated with *Chongryun* reportedly coordinated price negotiations and product specifications.[28] After the Japanese Coastguard exchanged fire with, and then sank, a North Korean spy ship in December 2001, officials discovered Japanese-made radar and other precision devices in the recovered vessel, raising questions about Japanese firms supplying equipment to the DPRK military.[29]

Some critics have recently suggested that Japanese politicians may intentionally allow cash remittances or high-tech material and contacts between *Chongryun* and North Korea to continue because of the politicians' own involvement with the Japanese underworld.[30] Such accusations have arisen more frequently since the end of Cold War, yet before that government agencies largely ignored the flow of money from Japan to North Korea, offering ferry links little surveillance and few cargo inspections as well as allowing *Chongryun* to maintain its tax-free diplomatic status for many years.[31] Indeed, President Park complained in 1969 to President Richard Nixon, saying that 'most of the equipment carried by North Korean guerrillas, who have infiltrated South Korea, such as their radios and shoes, are made in Japan. Although we made a protest against this, they continue to sell these items to make a profit.'[32] The record of conversations between Nixon and Park Chong Hee suggests that the ambivalent attitude of the Japanese government toward the North Korean regime under the two Koreas policy had contributed to an increase in the North Korean threat. In other words, Japan's ambivalent policy stance towards North Korea sometimes resulted in an accommodation policy towards the pro-Pyongyang Zainichi diaspora and virtually turned a blind eye to

CONCLUSION

Chongryun's role as Kim Il Sung's money pipeline. This stance continued until the end of 1990s, when Japan finally recognized the imminent North Korean threat and moved towards the US's previous proposal for Japan to destroy the pipeline of funds to North Korea. The perceived threat from North Korea converted Japan's fundamental long-term strategic trajectory and let the country move toward becoming a more normal military actor and a partner of the US symmetric alliance.

In this way, the Zainichi diaspora case study provides us with an important lesson – that exclusionary policies can play an indispensable role in the construction of identities that serve as the basis for mass mobilization. In the earlier stage of social mobilization, exclusionary policies allow the mass to lose connection with others, which ultimately becomes a basis for external involvement. Elites also played a crucial role in policy making, but it is important to note that an exclusionary policy which has been effective in order to deliberately marginalize a specific group will have unintended consequences, which might result in a heavy a cost for future generations, as the Zainichi diaspora case shows.

The lessons we can learn from history tell us that problems of identity can be seen to work in almost any historical setting. As Ernest Gellner shows, modernization is a typical example, in the sense that its process represents the transition from *Gemeinschaft* (community) to *Gesellschaft* (society), which constitutes a tremendous alienating process for those who have failed to catch up with the rapid transition. Hannah Arendt saw this kind of situation – in which the masses have 'lost their connection with others and become defined by their rootlessness' – as the indispensable precondition for totalitarian rule. Arendt defines this condition of the masses as *verlassenheit* – loneliness – or the state of being abandoned.[33] At this point we can observe the ways in which extremist politics began emerging in the twentieth century, driving alienated young people to become anarchists, Bolsheviks, fascists, or members of terrorist organizations.[34]

Exclusionary policies easily facilitate the mobilization of social movements against discrimination, at the same time consolidating the identity of ethnic minorities. However, it should be emphasized that social mobilization is not a by-product of something inherent to the cultural system that a particular culture has produced. Interestingly, Japanese Americans are a similar example of a minority that suffered severe exclusionary policy within a host country – the USA. On the eve of World War II, culturally vibrant and stable Nikkei communities (Japanese diaspora) were well rooted in the Pacific Northwest on both sides of the border. All this changed on 7 December 1941, when the Japanese military army attacked the United States naval base at Pearl Harbor, Hawaii, which led to the United States' entry into World War II, fighting against Japan. The end of March 1942 marked the start of the forced confinement of more than 110,000 Japanese Americans by the army. Two-thirds of those incarcerated were US citizens, while the rest were resident aliens who had been in the country for at least eight years but had been denied the right of naturalization. The American exclusionary policy, however, led to the construction of an identity within the ethnic community, which organized efforts to protect and promote their collective ethnic interests.[35] The whole process demonstrated just how quickly the eroded and neglected Japanese diaspora communities regained their vitality and fought back against the repressive measures imposed upon them by the USA in the postwar period.

It is also very obvious that the exclusionary policy emerged as a direct and immediate result of the abrupt change in interstate relations between Japan and America. The US entry into World War II not only put interstate relations with Japan onto an enemy footing but also allowed its policy toward enemy aliens to become an extremely strict exclusionary one. During World War II, the Japanese diaspora was considered as a potential enemy whose members might at any moment be called upon by their mother country to take up arms against America. In the postwar period, however – especially in 1952 – the McCarran–Walter Act removed race as a barrier to naturalization in the United States.[36] It should be also noted that the law was enacted just over one year after the US and Japan agreed to sign a security treaty in 1951, along with a subsidiary agreement that authorized US forces to use bases in Japan for Korean operations. By the end of the 1950s, the Japanese diaspora population was once again on the rise along the Pacific Coast. Although the Japanese diaspora community had already transformed its leadership from first to second generation by the 1950s, from the mid-1960s some third generation diaspora members became active in the American Civil Rights Movement, inspired by outrage at the experience of their elders during World War II and forming a driving force behind the redress movements in the United States. Ultimately, the

CONCLUSION

American exclusionary policy resulted in financial compensation as well as government apologies that began in 1988.[37]

This is the vital point that the Zainichi diaspora case study can help us learn. It underlines the recommendation that policy makers pursue policies of accommodation rather than exclusion in order to avoid paying the more expensive cost that exclusionary policies might bring in the long run. Given the fact that there are thousands of minority ethnic groups living in roughly 195 countries today – mostly concentrated in post imperial and post-colonial territories – it is hard to de-politicize these ethnic groups.[38] That is why the lessons from the history of the Zainichi diaspora underline an important recommendation that state policy makers should enhance interstate alliances through regional integration targeted not only for security and political purposes but also towards cultural or social cooperation in order to prevent exclusionary policies from taking hold. In so doing, they will need to prepare a definitive answer to the discussion that surrounds the question 'who are we?', and provide positive examples that define what it means to be a member of the larger community in order for diasporic populations not to find themselves gravitating towards a more powerful agent who expresses a greater certainty about their identity.

Notes

Introduction

1. John Lie, 2001. *Multiethnic Japan*. London: Harvard University Press, p. 171.
2. Erin Aeran Chung, 2010. *Immigration and Citizenship in Japan*. New York: Cambridge University Press.
3. The exact number is 381,645, according to 2012 statistics data provided by the Ministry of Justice in Japan: http://www.moj.go.jp/nyuukokukanri/kouhou/nyuukokukanri04_00030.html (accessed 18 March 2013).
4. Chikako Kashiwazaki, 2009. 'The Foreigner Category for Koreans in Japan,' in Sonia Ryang and John Lie (eds), *Diaspora without Homeland*. London: University of California Press, p. 145.
5. Apichai W. Shipper, 2008. *Fighting for Foreigners: Immigration and its Impact on Japanese Democracy*. Ithaca: Cornell University Press, p. 78.
6. Ibid.
7. Youngmi Lim, 2009. 'Korean Roots and Zainichi Routes,' in Sonia Ryang and John Lie (eds), *Diaspora without Homeland*. London: University of California Press, p. 93.
8. Chung, *Immigration and Citizenship*, p. 10.
9. Charles K. Armstrong, 2014. *The Koreas*. London: Routledge, p. 76.
10. Ibid., p. 80.
11. Enze Han, 2013. *Contestation and Adaptation*. Oxford: Oxford University Press.
12. Andrei Lankov, 2005. *Crisis in North Korea: The Failure of De-Stalinization, 1956*. Honolulu: University of Hawaii Press, quoted in Armstrong, *The Koreas*, p. 80.
13. Armstrong, *The Koreas*, p. 80.
14. Chung, *Immigration and Citizenship*, p. 75
15. Lori Watt, 2009. *When Empire Comes Home: Repatriation and Reintegration in Postwar Japan*. Cambridge: Harvard University Press, p. 93.
16. Ibid., p. 76.
17. Bumsoo Kim, 2006. 'From Exclusion to Inclusion? The Legal Treatment of "Foreigners" in Contemporary Japan,' *Immigrants & Minorities* 24(1): pp. 51–73.
18. Shipper, *Fighting for Foreigners*, p. 61.
19. For an exception, see Harris Mylonas, 2013. 'The Politics of Diaspora Management in the Republic of Korea,' The Asian Institute for Policy Studies, *Issue Brief No. 81*, 20 November 2013.
20. See, for instance: David Chapman, 2008. *Zainichi Korean Identity and Ethnicity*. London: Routledge; Chikako Kashiwazaki, 2000. 'The Politics of Legal Status', in Sonia Ryang (ed.), *Koreans in Japan*. London: Routledge; Yasunori Fukuoka, 1993. *Zainichi Kankoku-Chosenjin*. Tokyo: Chuo Koronsha; Sonia Ryang, 1997. *North Koreans in Japan*. Boulder, CO: Westview Press; and Erin Aeran Chung, 2010. *Immigration and Citizenship in Japan*. Cambridge: Cambridge University Press.
21. Seung-Young Kim, 2009. *American Diplomacy and Strategy toward Korea and Northeast Asia, 1882–1950 and After*. New York: Palgrave Macmillan.

NOTES TO PAGES 6–11

22. Kent E. Calder, 1988. 'Japanese Foreign Economic Policy Formation: Explaining the Reactive State,' *World Politics* 40(4): pp. 517–41.
23. Seung-Young Kim, 2013. 'Miki Takeo's Initiative on the Korean Question and U.S.-Japanese Diplomacy, 1974–1976,' *Journal of American-East Asian Relations*, 20: pp. 377–405, p. 377.
24. Christopher W. Hughes, 2009. 'Super-sizing the DPRK threat: Japan's evolving military posture and North Korea,' *Asian Survey* 49(2): pp. 291–311, p. 291.
25. Apichai W. Shipper, 2009. 'Nationalisms of and Against Zainichi Koreans in Japan,' *Asian Politics & Policy* 2(1): pp. 55–75; and Nicholas Eberstadt, 1996. 'Financial Transfers from Japan to North Korea: Estimating the Unreported Flows,' *Asian Survey* 36(5): pp. 523–42.
26. Fiona B. Adamson, 2002. 'Mobilizing for the Transformation of Home: Politicized Identities and Transnational Practices,' in Nadje Al-Ali and Khalid Koser (eds), *New Approaches to Migration? Transnational Communities and the Transformation of Home*. London: Routledge. See also Hazel Smith and Paul B. Stares, 2007. *Diasporas in Conflict: Peace-Makers or Peace-Wreckers?*. Tokyo; New York: United Nations University Press; and Khachig Tölölyan, 1991. 'The Nation-State and Its Others: In Lieu of a Preface,' *Diaspora: A Journal of Transnational Studies* 1(1): pp. 3–7.
27. Yossi Shain and Aharon Barth, 2003. 'Diasporas and International Relations Theory,' *International Organization* 57(03): pp. 449–79.
28. Smith and Stares, *Diasporas in Conflict*.
29. William Safran, 1991. 'Diasporas in Modern Societies: Myths of Homeland and Return,' *Diaspora: A Journal of Transnational Studies* 1(1): pp. 83–99.
30. Erin Aeran Chung, 2010. *Immigration and Citizenship in Japan*. New York: Cambridge University Press.
31. Rogers Brubaker, 1992. *Citizenship and Nationhood in France and Germany*. Cambridge, MA: Harvard University Press.
32. Lori Watt, 2009. *When Empire Comes Home: Repatriation and Reintegration in Postwar Japan*. Cambridge: Harvard University Press.
33. Rogers Brubaker, 1996. *Nationalism Reframed*. Cambridge: Cambridge University Press.
34. Harris Mylonas, 2012. *The Politics of Nation-Building, Making Co-Nationals, Refugees, and Minorities*. Cambridge: Cambridge University Press.
35. Mylonas also emphasizes the importance of the foreign policy goals of the host state in the process. If the host state is revisionist, then exclusionary policies are more likely toward an enemy backed group. If the host state retains the status quo then assimilationist policies become more likely. If the group is backed by an allied power, accommodationist policies are most likely. However, the particular condition that the Japan–South Korea–US triangle alliance provides needs the policy toward the enemy/allied backed group to be amended at some point. Threat perception is not sometimes correlated with foreign policy preference under strong superpower's security umbrella.
36. Nick Bryant, 2015. 'The Decline of US power?,' BBC News, 10 July 2015, http://www.bbc.co.uk/news/world-us-canada-33440287 (accessed 19 July 2015).
37. Victor D. Cha, 1999. *Alignment despite Antagonism: The United States-Korea-Japan Security Triangle*. Stanford, CA: Stanford University Press, p. 108.
38. Shain and Barth, 'Diasporas and International Relations Theory,' p. 450.
39. Mylonas, *The Politics of Nation-Building*, pp. 21–3.
40. David Chapman, 2008. *Zainichi Korean Identity and Ethnicity*. London: Routledge, p. 9.
41. Tong Myung Kim, 1988. *Zainichi chosenjin no daisan no michi*. Tokyo: Kaifusha.
42. Mika Ko, 2010. *Japanese Cinema and Otherness: Nationalism, Multiculturalism and the Problem of Japaneseness*. London: Routledge.
43. Benedict Anderson, 1991. *Imagined Communities: Reflections on the Origin and Spread of Nationalism*. London; New York: Verso.

44. Stuart Hall, 1990. 'Cultural Identity and Diaspora,' in J. Rutherford (ed.), *Identity: Community, Culture, Difference*. London: Lawrence and Wishart, pp. 222-37.
45. Ibid.
46. Paul Gilroy, 1993. *The Black Atlantic: Modernity and Double Consciousness*. Cambridge, MA: Harvard University Press.
47. Apichai W. Shipper, 2008. *Fighting for Foreigners: Immigration and Its Impact on Japanese Democracy*. Ithaca: Cornell University Press, p. 60.
48. Ibid., p. 61.
49. Chung, *Immigration and Citizenship in Japan*, p. 88.
50. Shipper, *Fighting for Foreigners*, p. 61.
51. Ibid.
52. Chung, *Immigration and Citizenship in Japan*, p. 87.
53. Shipper, *Fighting for Foreigners*, p. 74.
54. Sonia Ryang. 1997. *North Koreans in Japan*. Boulder, CO: Westview Press, p. 12.
55. Ibid., p. 107.
56. Ibid., p. 14.
57. Ibid., p. 124.
58. William Safran, 1991. 'Diasporas in Modern Societies: Myths of Homeland and Returns,' *Diaspora: A Journal of Transnational Studies* 1(1): pp. 83-99.
59. James Clifford, 1994. 'Diasporas,' *Cultural Anthropology* 9(3): pp. 302-8.
60. Tölöyan Kaching, 1991. 'Rethinking Diasporas: Stateless Power in the Transnational World,' *Diaspora: A Journal of Transnational Studies* 5(1): pp. 3-36.
61. For example, see Duncan Green, 2015. 'Migrant remittances are even more amazing that we thought,' 30 January 2015, He describes in terms of economy, 'Remittances from migrants to developing countries are now running at some three times the volume of aid, and barely faltered during the 2008-9 financial crisis ... Remittances are associated with more stable domestic consumption growth', http://oxfamblogs.org/fp2p/migrant-remittances-are-even-more-amazing-that-we-thought/ (accessed 30 January 2015).
62. Khachig Tölölyan, 1991. 'The Nation-State and Its Others: In Lieu of a Preface,' *Diaspora: A Journal of Transnational Studies* 1(1): pp. 3-7.
63. Appadurai Arjun, 1996. *Modernity at Large: Cultural Dimensions of Globalization*. Minneapolis: University of Minnesota Press.
64. Ibid., p. 494.
65. Zlatko Skrbis, 2007. 'The mobilized Croatian diaspora: Its role in homeland politics and war,' in Hazel Smith and Paul States (eds), *Diasporas in Conflict: Peace-Makers or Peace-Wreckers*. Tokyo: United Nations University.
66. Rogers Brubaker, 1996. *Nationalism Reframed*. Cambridge: Cambridge University Press.
67. Ibid.
68. Leo T.S. Ching, 2001. *Becoming 'Japanese': Colonial Taiwan and the Politics of Identity Formation*. London: University of California Press.
69. See, for instance, Gabriel Sheffer, 2003. *Diaspora Politics: At Home Abroad*. New York: Cambridge University Press; and Yossi Shain, 2007. *Kinship and Diasporas in International Affairs*. Ann Arbor: University of Michigan Press.
70. In the Westphalia system, the national interests and goals of states (and later nation states) were widely assumed to go beyond those of any citizen or any ruler. States became the primary institutional agents in an interstate system of relations.
71. Mylonas, *The Politics of Nation-Building*, pp. 53-5.
72. Michael Weiner, 1997. 'The invention of identity: "Self" and "Other" in pre-war Japan,' in Michael Weiner (eds), *Japan's Minorities: The Illusion of Homogeneity*. London; New York: Routledge, p. 1.

73. A.D. Smith, 1999. *Myths and Memories of the Nation*. Oxford: Oxford University Press.
74. Ibid.
75. Fiona B. Adamson, 2012. 'Constructing the Diaspora: Diaspora Identity Politics and Transnational Social Movements,' in Terrence Lyons and Peter Mandaville (eds), *Politics from Afar: Transnational Diasporas and Networks*. London: Hurst & Co., p. 32.
76. Chapman, *Zainichi Korean Identity and Ethnicity*; and Shipper, *Fighting for Foreigners*.
77. Chung, *Immigration and Citizenship in Japan*, p. 51.
78. Adamson, 'Constructing the Diaspora', p. 33.
79. Andrei Lankov, 2010. 'Forgotten people: the Koreans of Sakhalin Island in 1945–1991,' www.nkeconwatch.com/nk-uploads/Lankov-Sakhalin-2010.pdf (accessed 24 May 2015).
80. Erin K. Jenne, 2008. *Group Demands as Bargaining Positions: Signals, Cues and Minority Mobilization in East Central Europe*. UMI Number: 9995233, p. 13
81. Alexander Wendt, 1994. 'Collective Identity Formation and International State,' *The American Political Science Review* 88(2): pp. 384–96.
82. Roland Bleiker, 2005. *Divided Korea: Toward a Culture of Reconciliation*. London: University of Minnesota Press.
83. Paul Brass, 1991. *Ethnicity and Nationalism: Theory and Comparison*. London: Sage Publications, p. 74.
84. Alex Delmar-Morgan and Peter Osborne, 2014. 'Why is the Muslim charity Interpal being blacklisted as a terrorist organization?,' *The Telegraph*, 26 November 2014, http://www.telegraph.co.uk/news/religion/11255294/Why-is-the-Muslim-charity-Interpal-being-blacklisted-as-a-terrorist-organisation.html. 'Islamic charities in the UK find themselves in choppy waters as they face extraordinary scrutiny and pressure. In recent weeks, as David Cameron awarded the Charity Commission extra powers to investigate extremism, this has escalated.'
85. George Barany, 1974. 'Magyar Jew or Jewish Magyar?: Reflections on the question of assimilation,' in *Jews and Non-Jews in Eastern Europe, 1918–1945*. New York: Wiley and Sons.
86. Karl W. Deutsch, 1968. 'The Trend of European Nationalism – the Language Aspect,' in Joshua A. Fishman (eds), *Readings in the Sociology of Language*. The Hague: Mouton.
87. Ibid., p. 77.
88. Brass, *Ethnicity and Nationalism*, p. 30.
89. Ibid., p. 45
90. Fredrik Barth, 1998. *Ethnic Groups and Boundaries: The Social Organization of Culture Difference*. Prospect Heights, IL: Waveland Press.
91. Suny Ronald Grigor, 2001. 'Constructing Primordialism: Old Histories for New Nations,' *The Journal of Modern History* 73(4): pp. 862–96.
92. Hugh Seton-Watson, 1977. *Nations and States: An Enquiry into the Origins of Nations and the Politics of Nationalism*. London: Methuen.
93. Chung, *Immigration and Citizenship in Japan*.
94. Armstrong, *The Koreas*, p. 76.
95. Ibid., p. 77
96. Ernest Gellner, 2006. *Nations and Nationalism*. 2nd edn. Malden, MA: Blackwell Publisher.
97. Ibid.
98. Mark R. Peattie, 1984. 'Japanese Attitudes toward Colonialism,' in Ramon H. Myers and Mark R. Peattie (eds), *The Japanese Colonial Empire, 1895–1945*. Princeton: Princeton University Press, pp. 80–127.

99. Ibid., p. 25.
100. Lori Watt, 2009. *When Empire Comes Home*. Cambridge: Harvard University Press.
101. Michael Hechter, 1975. *Imagined Colonialism: The Celtic Fringe in British National Development, 1536–1966*. Berkeley: University of California Press, pp. 60–4.
102. David Newman, 2000. 'Citizenship, Identity and Location,' in Klaus Dodds and David Atkinson (eds), *Geopolitical Traditions: A Century of Geopolitical Thought*. London: Routledge.
103. Rogers Brubaker, 2005. 'The "diaspora" diaspora,' *Ethnic and Racial Studies* 28(1): pp. 1–19.
104. Ibid. p. 12.
105. Primordial perspectives consider diasporas as natural entities that emerge simply out of territorial boundaries, but Brubaker explores the idea that diasporas represent themselves as symbolic existence and discursively constructed.
106. Brubaker, 'The "diaspora" diaspora', p. 12.
107. Miles Kahler, 1997. 'Empires, Neo-Empires and Political Change,' in K. Dawisha and B. Parrott (eds), *The End of Empire: The Transformation of the USSR in Comparative Perspective*. Armonk, NY: M. E. Sharpe, p. 288.
108. Suny Ronald Grigor, 2001. 'The Empire Strikes Out,' in Ronald Grigor Suny and Terry Martin (eds), *A State of Nations: Empire and Nation-making in the Age of Lenin and Stalin*. Oxford; New York: Oxford University Press, p. 35.
109. John Darwin, 1988. *Britain and Decolonization*. New York: St. Martin's Press, p.16.
110. Kahler, 'Empires, Neo-Empires and Political Change', p. 288.
111. Ibid.
112. Ibid.
113. Jennifer Lind, 2008. *Sorry States*. Ithaca; London: Cornell University Press.
114. Victor D. Cha, 1999. *Alignment despite Antagonism: The United States-Korea-Japan Security Triangle*. Stanford, CA: Stanford University Press.
115. Ibid.
116. Fiona B. Adamson, 2013. 'Mechanisms of diaspora mobilization and the Transnationalization of Civil War,' in Jeffrey T. Checkel (ed.), *Transnational Dynamics of Civil War*. Cambridge: Cambridge University Press, p. 71.

Chapter 1: Alliance Cohesion, Diaspora and Nation-Building Policies

1. Pratt Cranford (ed.), 1990. *Middle Power Internationalism: The North-South Dimension*. Kingston and Buffalo: McGill-Queen's University Press.
2. Mylonas, *The Politics of Nation-Building*, p. 25.
3. Robert Cox, 1989. 'Middlepowermanship, Japan and Future World Order,' *International Journal* 44(4): pp. 823–62, 825.
4. David Kang, 2007. *China Rising*. New York: Columbia University Press. 'While Korea and Japan had constituted the inner core of the Chinese-dominated regional system over history, the universal empire presided over by the Son of Heaven (the Chinese Emperor) had risen and was maintained in East Asia until the middle of the nineteenth century. After the end of World War II, the United States continually has taken a leadership role considered vital for the region's security.'
5. Robert Cox, 'Middlepowermanship', p. 827. Also see, Pratt Cranford (ed.), 1990. *Middle Power Internationalism: The North-South Dimension*. Kingston and Buffalo: McGill-Queen's University Press.

6. Cranford, *Middle Power Internationalism*, pp. 5–7.
7. Ibid., p. 5.
8. Ibid., pp. 23–4.
9. Mylonas adds footnotes as follows: 'This is especially true for small powers in a bipolar or unipolar international system,' so it was true for Japanese ruling elites as well, especially during the Cold War era.
10. Cranford, *Middle Power Internationalism*, pp. 26–7.
11. Ibid., pp. 21–3.
12. Mylonas, *The Politics of Nation-Building*, pp. 17–49.
13. Jennifer Lind, 2008. *Sorry States*. Ithaca; London: Cornell University Press.
14. E. Takemae, R. Ricketts, S. Swann and J.W. Dower, 2002. *The Allied Occupation of Japan*. New York; London: Continuum, p. 495.
15. Takemae et al., *The Allied Occupation of Japan*, p. 510.
16. Ibid.
17. Chikako Kashiwazaki, 2000. 'The Politics of Legal Status,' in Sonia Ryang (ed.), *Koreans in Japan: Critical Voices from the Margin*. London: Routledge.
18. Mylonas, *The Politics of Nation-Building*, p. 30.
19. This evidence is consistent with Mylonas's main argument. Japan seemed to treat the Zainichi as two distinct groups, excluding the North Korean sympathizers, but repatriations were not necessarily driven only by the sympathy toward North Korea but rather by the hope of escaping from the misery of life in Japan. From the view of Japanese government, the primary policy was to eliminate the presence of Koreans in Japan, who were considered to be a source of criminality and political unrest regardless of sympathizing either with the South or the North, but Rhee's neglect of the welfare of Koreans in Japan and his scant concern for those who might wish to return to their homeland allowed the Japanese government to promote the plan to repatriate Zainichi Koreans to North Korea. See Changsoo Lee and George A. De Vos, 1981. *Koreans in Japan: Ethnic Conflict and Accommodation*. Berkeley: University of California Press, pp. 91–109.
20. Tessa Morris-Suzuki, 2007. *Exodus to North Korea: Shadows from Japan's Cold War*. Lanham, MD: Rowman & Littlefield Publishers.
21. Yossi Shain, 2007. *Kinship & Diasporas in International Affairs*. Ann Arbor: University of Michigan Press, p. 143.
22. The repatriation movement of Zainichi diaspora shows that the Japanese government was not interested in helping Koreans make a decision about citizenship or in helping them repatriate to a Korean government of their own choice. The South Korean government showed little concern about the Zainichi diaspora although the North Korean government saw its political advantage. Nothing resolved the problem of identity being faced by the Zainichi diaspora. See Changsoo and De Vos. *Koreans in Japan*, p. 109.
23. Benjamin Miller, 2007. *States, Nations, and the Great Powers: The Sources of Regional War and Peace*. Cambridge; New York: Cambridge University Press, p. 108.
24. Ibid.
25. Mylonas, *The Politics of Nation-Building*, p. 30.
26. Stephen M. Walt, 1997. 'Why Alliances Endure or Collapse,' *Survival* 39(1): p. 157.
27. Mylonas, *The Politics of Nation-Building*, p. 110.
28. James D. Morrow, 1991. 'Alliances and Asymmetry: An Alternative to the Capability Aggregation Model of Alliances,' *American Journal of Political Science* 35(4): pp. 904–33, p. 915.
29. Ibid., p. 914.
30. James D. Morrow, 1994. *Game Theory for Political Scientists*. Princeton, NJ: Princeton University Press.

31. James D. Morrow, 1986. 'A spatial model of international conflict,' *American Political Science Review* 80(4): pp. 1131-50.
32. Glenn H. Snyder, 1990. 'Alliance Theory: A Neorealist First Cut,' *International Organization* 45(1): pp. 121-42.
33. Brubaker, 'The "diaspora" diaspora', pp. 1-19.
34. Yossi Shain and Aharon Barth, 2003. 'Diasporas and International Relations Theory,' *International Organization* 57(3): pp. 449-79, p. 451.
35. Glenn H. Snyder, 1984. 'The Security Dilemma in Alliance Politics,' *World Politics* 36(4): pp. 461-96; and Cha, *Alignment despite Antagonism*, p. 43.
36. Ryang, *North Koreans in Japan*, p. 125.
37. Ibid., p. 124
38. Lee and De Vos, *Koreans in Japan*, p. 118.
39. Ibid., p. 451.
40. Shipper, 'Nationalisms of and Against Zainichi Koreans', p. 59.
41. Ibid., p. 60.
42. Erin K. Jenne, 2007. *Ethnic Bargaining: The Paradox of Minority Empowerment*. Ithaca: Cornell University Press.
43. Cha, *Alignment despite Antagonism*.
44. Brett Ashley Leeds and Savun Burcu, 2007. 'Terminating Alliances: Why Do States Abrogate Agreements?,' *The Journal of Politics* 69(4): pp. 1118-32.
45. Alexander L. George and Andrew Bennett, 2005. *Case Studies and Theory Development in the Social Sciences*. Cambridge: MIT Press, p. 223.
46. Charles Tilly, 1997. 'Means and Ends of Comparison in Macrosociology,' *Comparative Social Research* 16: pp. 43-53.
47. George and Bennett, *Case Studies and Theory Development*, p. 205.
48. Ibid.
49. The Foreign Relations of the United States, National Archives and Records Administration (NARA) The Library of Congress, National Archives, Nixon Presidential Materials, The Digital National Security Archive (DNSA), National Archives, Nixon Presidential Materials, Central Intelligence Agency, Joint Chiefs of Staff, 1945.
50. *Yomiuri* Newspaper; *Asahi* Newspaper; *Japan Times*; and BBC World News.
51. Japan Red Cross Society; Ministry of Justice in Japan; and Japanese Ministry of Foreign Affairs.

Chapter 2: The Zainichi Diaspora: From the Shadow of Japan's Colonial Legacy

1. Toshiyuki Tamura, 2003. 'The Status and Role of Ethnic Koreans in the Japanese Economy,' in C. Fred Bergsten and In-bŏm Ch'oe (eds), *The Korean Diaspora in the World Economy*. Washington DC: Institute for International Economics, pp. 77-9.
2. Ibid., p. 79.
3. Ibid., p. 80.
4. Michael Weiner, 1989. *The Origins of the Korean Community in Japan: 1910-1923*. Manchester: Manchester University Press, p. 74.
5. Chong-Sik Lee, 1985. *Japan and Korea: The Political Dimension*. Stanford: Hoover Institution Press.
6. Ibid., p. 15.

7. In-bŏm Ch'oe, 2003. 'Korean Diaspora in the Making: Its Current Status and Impact on the Korean Economy,' in Bergsten and Ch'oe (eds), *The Korean Diaspora in the World Economy*, p. 17.
8. Ibid., p. 15.
9. Helder De Schutter and Lea Ypi, 2015. 'The British Academy Brian Barry Prize Essay: Mandatory Citizenship for Immigrants,' *British Journal of Political Science* 45(2): pp. 235–51: 'According to current international law only the state of origin is compelled to accept citizens into its national territory ... Naturalization, however, is not granted unconditionally in their state of residence.'
10. Tessa Morris-Suzuki, 2010. *Borderline Japan: Foreigners and Frontier Controls in the Postwar Era*. New York: Cambridge University Press, p. 7.
11. Historically, the imperial powers often forced weaker states to grant extraterritorial rights to their citizens who were not diplomats – including soldiers, traders, Christian missionaries, etc. This was most famously the case in East Asia during the nineteenth century, where China and Japan were not formally colonized, but were subjugated to an extent by the western powers.
12. Mark R. Peattie, 1984. *Introduction*, in Ramon H. Myers, Mark R. Peattie and Ching-chih Chen (eds), *The Japanese Colonial Empire, 1895–1945*. Princeton, NJ: Princeton University Press.
13. W.G. Beasley, 1987. *Japanese Imperialism: 1894–1945*. Oxford: Clarendon Press.
14. Peattie et al., *The Japanese Colonial Empire*, p. 8.
15. Eric Hobsbawm, 1997. *The Age of Empire, 1875–1914*. London: Abacus.
16. Weiner, *The Origins of the Korean Community*, p. 7.
17. Mark Caprio, 2009. *Japanese Assimilation Policies in Colonial Korea, 1910–1945*. Seattle: University of Washington Press. Caprio argues that Japanese were inspired by British, Prussian, and French efforts in their peripheral territories, rather than these states' efforts in their external possessions.
18. Edward I-te-Chen, 1984. 'The Attempt to Integrate the Empire, 1895–1945,' in Ramon H. Myers and Mark R. Pettie (eds), *The Japanese Colonial Empire: 1895–1945*. Princeton: Princeton University Press, p. 243.
19. Myers et al., *The Japanese Colonial Empire*, p. 17.
20. As cited in Weiner, *The Origins of the Korean Community*, p. 20.
21. Ibid.
22. Dae-yeol Ku, 1985. *Korea Under Colonialism: The March First Movement and Anglo-Japanese Relations*. Seoul: Seoul Computer Press.
23. I-te-Chen, 'The Attempt to Integrate the Empire', p. 250.
24. Ibid.
25. Ibid.
26. Ibid., p. 251.
27. Mark Caprio, 2009. *Japanese Assimilation Policies in Colonial Korea, 1910–1945*. Seattle: University of Washington Press.
28. Marie Seong-Hak Kim, 2012. *Law and Custom in Korea: Comparative Legal History*. Cambridge: Cambridge University Press.
29. I-te-Chen, 'The Attempt to Integrate the Empire', pp. 249, 274.
30. Kim, *Law and Custom in Korea*, p. 251.
31. Ramon H. Myers, 1984. 'Post World War II Japanese Historiography of Japan's Formal Colonial Empire,' in Peattie et al., *The Japanese Colonial Empire*, p. 460.
32. Peattie et al., *The Japanese Colonial Empire*, p. 29.
33. Weiner, *The Origins of the Korean Community*, p. 33.

34. Sang-ch'ŏl Sŏ, 1978. *Growth and Structural Changes in the Korean Economy, 1910–1940.* Cambridge: Harvard University Press.
35. Yangban control over rural life was furthered by the maintenance of aristocratic privilege and the virtual monopolization of political power by the members of this class. See Yunshik Chang, 1971. 'Colonization as Planned Change: The Korean Case,' *Modern Asian Studies* 5(2): pp. 161–86.
36. Sang-Chul Suh, 1978. *Growth and Structural Changes in the Korean Economy 1910–1940.* London: Harvard University Press.
37. Bruce Cumings, 1981. *The Origin of the Korean War: Liberation and the Emergence of Separate Regimes, 1945–1947.* Princeton: Princeton University Press.
38. Bruce Cumings, 1984. 'The Legacy of Japanese Colonialism in Korea', in Myers and Peattie (eds), *The Japanese Colonial Empire: 1895–1945*, p. 490.
39. Weiner, *The Origins of the Korean Community*, p. 42.
40. Bruce F. Johnson, 1953. *Japanese Food Management in World War II.* Stanford: Stanford University Press.
41. For about the first five years after the annexation, the number of Koreans migrating to southern Manchuria and Chientao region exceeded those moving to Japan. The demands for Korean labour in Japan began from 1916. See, Setsure Tsurushima, 1978. 'Korean Immigrants in Kando in the 1920s,' *The Kansai University Review of Economics and Business* 7: pp. 48–52.
42. Weiner, *The Origins of the Korean Community*, p. 43; also see Figure 2.2 in Chapter 1, we can find the increased Korean population in Japan since around 1920.
43. Ibid., p. 36.
44. Ibid., p. 37.
45. Ibid.
46. Hildi Kang, 2001. *Under the Black Umbrella: Voices from Colonial Korea, 1910–1945.* Ithaca: Cornell University Press.
47. Saito's first major address on 3 September 1919, cited in Weiner, *The Origins of the Korean Community*, p. 35.
48. Pak Soon-Yong and Keumjoong Hwang, 2011. 'Assimilation and Segregation of Imperial Subjects: "educating" the Colonized during the 1910–1945 Japanese Colonial Rule of Korea,' *Paedagogica Historica* 47(3): pp. 377–97, p. 386.
49. Ibid., p. 387.
50. Ibid., p. 388.
51. Peter Duus, 1996. 'Introduction,' in Peter Duus, Ramon Hawley Myers, Mark R. Peattie, and Wan-yao Chou (eds), *The Japanese Wartime Empire, 1931–1945.* Princeton, NJ: Princeton University Press.
52. Ibid.
53. Ibid.
54. Carter J. Eckert, 1996. 'Total War, Industrialization and Social Change in Late Colonial Korea,' in Peter Duus, Ramon H. Myers, and Mark R. Peattie (eds), *The Japanese Wartime Empire 1931–1945.* Princeton: Princeton University Press.
55. Lee, *Japan and Korea*, p. 7.
56. Ibid.
57. *Chosen Shotokufu kanpo kogai* [Chosen Shotokufu newsletter, extra edition] 1938:2.
58. Yusaku Kodawau, 1938. *Kokoku shinmin taruno chigaku no dettei suru chosen kyoikurei no gaisetsu yori*, in *Bunkyo no Chosen*, March Edition. Quoted in Soon-Yong and Hwang, 'Assimilation and Segregation,' p. 390. In other words, *Kokoku shinmin ka* means a policy to transform Koreans into the imperial nation's subjects. See Wan-yao Chou et al. (eds), *The Japanese Wartime Empire*; Chou's main argument is that the *Kokoku shinmin*

movement in Korea was all about ethnic conversion, which was different from *Kominka* policy in Taiwan. p. 40 also see p. 43 '... in the case of Korea, the concept of "the nation" was clearly emphasized in the oath ... this seems to suggest that the colonial government in Korea had to compete, in its endeavour to transform Koreans into Japanese, with the historical fact that Korea had once been an independent nation.'
59. Wan-yao Chou, 1996. 'The Kominka Movement in Taiwan and Korea: Comparisons and Interpretation' in Duus, Myers and Peattie (eds), *The Japanese Wartime Empire*.
60. Noboru Kamihara,1939. *Komin shinmin ikusei no genjo* [*The Present State of Bringing Up Imperial Subjects*], in *Chosen*, pp. 39–48.
61. *Chosen Sotokufu* [Korean Government-General], 1932, *Tokei nempo*, Seoul.
62. Naoki Mizuno, 2004. 'Chosenjin no namae to shokuminchi-shugi' ['Names of Koreans and Colonial Rule'], in Kyoto Naoki Mizuno (ed.), *Seikatsu no naka no shokuminchi-shugi, Jimbun Shoin*, pp. 35–77.
63. Quoted in Bruce Cumings, 1984. 'The Legacy of Japanese Colonialism in Korea,' in Myers and Pettie (eds), *The Japanese Colonial Empire*. Chong-Sik Lee also argues in his literature, *Japan and Korea, (1985)*, approximately one out of every two households sent a male worker to the Japanese mines and factories in Korea.
64. Senji Tsuboe, 1965, Zai Nihon Chosenjin no gaikyo, Tokyo.
65. Wan-yao Chou, 1996. 'The Kominka Movement in Taiwan and Korea: Comparisons and Interpretation,' in Ramon Hawley Myers, Mark R. Peattie and Ching-chih Chen (eds), *The Japanese Wartime Empire, 1931–1945*. Princeton, NJ: Princeton University Press, p. 65.
66. Ibid., p. 64.
67. Eckert, 'Total War, Industrialization and Social Change', p. 17.
68. Ibid., p. 25.
69. Bruce Cumings, 1997. *Korea's Place in the Sun: A Modern History*. London: W.W. Norton & Company.
70. Chou, 'The Kominka Movement in Taiwan', p. 67. Also see, George H. Kerr, 1966. *Formosa Betrayed*. London: Eyre and Spottiswoode. During the immediate postwar period, the Kuomintang (KMT) administration on Taiwan was repressive and extremely corrupt compared with the previous Japanese rule, leading to local discontent. The best-known and most serious conflict was the anti-mainlander violence between the Taiwanese and the Nationalist government, which flared up on 28 February 1947.
71. Cumings, *Korea's Place in the Sun*. Also see I-te-Chen, 'The Attempt to Integrate the Empire': 'Given the enthusiasm generated by the League for the Establishment of a Taiwan ... even in the hostile environment of that colony ... it would have had far greater popular appeal in competing with Korean aspiration for outright independence than trying to integrate the colony by representation in the Imperial Diet,' p. 274.
72. Cumings, *Korea's Place in the Sun*, p. 158.
73. Ibid.
74. Peattie, *Introduction*, p. 46.
75. Sonia Ryang, 2008. *Writing Selves in Diaspora*. New York: Lexington Books.
76. M.C. Sandusky, 1983. *America's Parallel*. Alexandria: Old Dominion Press.
77. Ibid.
78. Suh Dae-Sook, 1967. *The Korean Communist Movement, 1918–1948*. Princeton: Studies of the East Asian Institute at Columbia.
79. Won-sul Lee, 1982. *The United States and the Division of Korea, 1945*. Seoul: Kyunghee University Press.
80. Hyung Kook Kim, 1995. *The Division of Korea and the Alliance Making Process: Internationalization of Internal Conflict and Internalization of International Struggle, 1945–1948*. London: University Press of America.

81. John Lewis Gaddis, 1972. *The US and the Origins of the Cold War*. New York: Columbia University Press.
82. Bruce Cumings, 1981. *The Origins of the Korean War: Liberation and the Emergence of Separate Regimes, 1945-1947*. Princeton: Princeton University Press.
83. Peter Lowe, 1986. *The Origins of the Korean War*. London: Longman Group Limited.
84. Masao Okonigi, 1977. 'The Domestic Roots of the Korean War', in Yonosuke Nagai and Akira Iriye (eds), *The Origins of the Cold War in Asia*. Tokyo: Tokyo University Press.
85. Ibid., p. 300.
86. T. Tamura, 1999. *Korean Population in Japan 1910-1945*. (1)-(3) Keizai-to Keizaigakusha.
87. E.W. Wagner, 1951. *Korean Minority in Japan 1904-50*. New York: Institute of Pacific Relations.

Chapter 3: No Alliance and a Strong Historical Legacy: Exclusionary Policies towards the Zainichi in the Post-World War II Era (1945-64)

1. George Liska, 1962. *Nations in Alliance: The Limits of Interdependence*. Baltimore, MD: Johns Hopkins University Press, p. 3.
2. Morrow, 'Alliances and Asymmetry'.
3. Stephen M. Walt, 1997. 'Why Alliances Endure or Collapse,' *Survival* 39(1): pp. 156-79, p. 157, see also note 3.
4. Jae Jeok Park, 2011. 'The US-led Alliances in the Asia-Pacific: Hedge against Potential Threats or an Undesirable Multilateral Security Order?,' *The Pacific Review* 24(2): pp. 137-58.
5. Victor Cha, 2011. '"Rhee-strait": The Origins of the U.S.-ROK Alliance,' *International Journal of Korean Studies* 15(1): pp. 1-15.
6. Ibid.
7. Cumings, *Korea's Place in the Sun*, p. 159.
8. Erez Manela, 2007. *The Wilsonian Moment: Self-Determination and the International Origins of Anticolonial Nationalism*. New York and Oxford: Oxford University Press.
9. Soon Sung Cho, 1967. *Korea in World Politics, 1940-1950*. Berkeley: University of California Press.
10. Haruki Wada, 2014. *The Korean War: An International History*. New York: Roman & Littlefield, p. 2.
11. Japanese Ministry of Foreign Affairs, 1954. *Komura Gaiko Shi (History of Komura's Diplomacy)*. Tokyo, vol. 29, pp. 812-13, *Nihon Gaiko Bunsho*.
12. Japanese Ministry of Foreign Affairs, 1953. *Komura Gaiko Shi (History of Komura's Diplomacy)*. Tokyo, vol. 1, p. 97, *Nihon Gaiko Bunsho*. See also Soon Sung Cho, 1963. *Korea in World Politics, 1940-1950*. Berkeley: University of California Press, p. 50: 'September 1903, a few months before the outbreak of hostilities, Russian minister Rosen sought Japanese accord for an agreement making a neutral demilitarized zone in Korea north of the 39th parallel while recognizing Japan's special interests in the south. The Japanese counter-proposed a ten mile wide demilitarized zone along the boundary between Manchuria and Korea, that is, five miles to the north of the boundary and five miles to the south of it. The proposal was turned down and the war between the two nations (1904-5) finally settled the issue. Russian influence was ended by defeat. So the division line of the 39th parallel would have meant that the US could occupy Dairen in Manchuria.'
13. Chae-Jin Lee, 2006. *A Troubled Peace: US Policy and the Two Koreas*. Baltimore: John Hopkins University Press, p. 20.

14. Ibid.
15. Ibid.
16. Anthony Farrar-Hockley, 1990. *The British Part in the Korean War*. London: HMSO.
17. Gerald L. Curtis and Sŭng-jŭ Han, 1983. *The U.S.-South Korean Alliance: Evolving Patterns in Security Relations*. Lexington, MA: Lexington Books, p. 2.
18. Masao Okonogi, 1985. 'A Japanese Perspective on Korea-Japan Relations,' in Chin-Wee Chung, Ky-Moon Ohm, Suk-Ryil Yu and Dal-Joong Chnag (eds), *Korea and Japan in World Politics*. Seoul: Korean Association of International Relations, p. 19.
19. James William Morley, 1983. 'The Dynamics of the Korean Connection', in Curtis and Han (eds), *The U.S.-South Korean Alliance*, p. 15.
20. Gregg Brazinsky, 2007. *Nation Building in South Korea: Koreans, Americans, and the Making of a Democracy*. Chapel Hill: University of North Carolina Press, p. 13.
21. Morley, 'The Dynamics of the Korean Connection', p. 10.
22. William Whitney Stueck, 1981. *The Road to Confrontation: American Policy Toward China and Korea, 1947-1950*. Chapel Hill: University of North Carolina Press, pp. 20-1.
23. Stueck, *The Road to Confrontation*, p. 21.
24. The final running-mate of President Franklin D. Roosevelt in 1944, Truman succeeded to the presidency on 12 April 1945, when Roosevelt died after months of declining health.
25. Stueck, *The Road to Confrontation*, p. 21.
26. Ibid., p. 24.
27. Ibid., p. 25.
28. Ibid., p. 29.
29. John Lewis Gaddis, 1977. 'Korea in American Politics, Strategy, and Diplomacy, 1945-50,' in Yonosuke Nagai and Akira Iriye (eds), *The Origins of the Cold War in Asia*. Tokyo: University of Tokyo Press.
30. Stueck, *The Road to Confrontation*, p. 79.
31. Brazinsky, *Nation Building in South Korea*, p. 14.
32. Ibid., p. 15.
33. Stueck, *The Road to Confrontation*, p. 76.
34. Ibid., p. 80.
35. Ibid.
36. Foreign Relations of the United States, 1947. *The Secretary of War (Patterson) to the Acting Secretary of State*. FRUS: The Far East, Vol. VI, p. 626, 740.00119 Control (Korea)/4-447.
37. Adam B. Ulam, 1968. *Expansion and Coexistence: The History of Soviet Foreign Policy, 1917-1967*. New York: Praeger.
38. Morley, 'The Dynamics of the Korean Connection', p. 12.
39. Foreign Relations of the United States, 1947. The Secretary of War (Patterson) to the Acting Secretary of State. FRUS, The Far East Vol. VI, pp. 626-7, 740.00119 Control (Korea)/4-447.
40. Soon Sung Cho, 1967. *Korea in World Politics, 1940-1950*. Berkeley: University of California Press.
41. Chae-Jin Lee and Hideo Sato, 1982. *U.S. Policy Toward Japan and Korea: A Changing Influence Relationship*. New York: Praeger Publications.
42. Nikita Khrushchev, 1974. *Khrushchev Remembers: The Last Testament*, trans. and ed. Strobe Talbott. London: Andre Deutsch.
43. Morley, 'The Dynamics of the Korean Connection', p. 21.
44. Lee, *A Troubled Peace*, p. 23.
45. Stueck, *The Road to Confrontation*, p. 90.

46. Lee, *A Troubled Peace*, p. 24.
47. Ibid.
48. Ibid., p. 23.
49. Ibid., p. 25.
50. Ibid., p. 24.
51. Stueck, *The Road to Confrontation*, p. 10.
52. Thomas P. Bernstein and Andrew J. Nathan, 1983. 'The Soviet Union, China and Korea,' in Gerald L. Curtis and Sung-joo Jan (eds), *The US-South Korean Alliance: Evolving Patterns in Security Relations*. Lexington, MA: Lexington Books, pp. 89–91.
53. Allen S. Whiting, 1968. *China Crosses the Yalu: The Decision to Enter the Korean War*. Stanford: Stanford University Press.
54. Bernstein and Nathan, 'The Soviet Union, China and Korea', p. 94.
55. Ibid., p. 95.
56. Ibid.
57. Ibid.
58. Stephen S. Kaplan, 1981. *Diplomacy of Power: Soviet Armed Forces as a Political Instrument*. Washington, DC: Brookings Institution.
59. Lee, *A Troubled Peace*, p. 31.
60. Foreign Relations of the United States, 1951, *Memorandum Containing the Sections Dealing with Korea From NSO 48/5*, 17 May 1951, FRUS, Vol. VII, Part 1, pp. 439–42, 795-00/5-1751.
61. Morley, 'The Dynamics of the Korean Connection', p. 17.
62. Ibid.
63. Lee, *A Troubled Peace*, p. 38.
64. Byung-joon Ahn, 1983. 'The United States and Korean-Japanese Relations,' in Gerald L. Curtis and Sŭng-jŭ Han (eds), *The U.S.-South Korean Alliance: Evolving Patterns in Security Relations*. Lexington, MA: Lexington Books, p. 130.
65. Ibid., p. 129.
66. Michael Schaller, 1997. *Altered States: The United States and Japan since the Occupation*. New York: Oxford University Press.
67. Ibid., Article IX states: 'Aspiring sincerely to an international peace based on justice and order, the Japanese people forever renounce war as a sovereign right of the nation and the threat or use of force as a means of settling international disputes.'
68. J.W. Dower, 1979. *Empire and Aftermath: Yoshida Shigeru and the Japanese Experience, 1878–1954*. Cambridge, MA: Harvard University Press.
69. Shigeru Yoshida, 2007. *Yoshida Shigeru: Last Meiji Man*. Lanham: Rowman & Littlefield.
70. John Welfield, 1988. *An Empire in Eclipse: Japan in the Postwar American Alliance System: a Study in the Interaction of Domestic Politics and Foreign Policy*. London; Atlantic Highlands, NJ: Athlone Press, p. 41.
71. Ibid., p. 46.
72. Ibid., p. 50.
73. Ibid., p. 51.
74. Schaller, *Altered States*, p. 36.
75. Foreign Relations of the United States, 1951. *The Consultant to the Secretary (Dulles) to the Secretary of State*, 10 February 1951, FRUS, Vol. VI, pp. 874–80.
76. Schaller, *Altered States*, p. 35. See also, Foreign Relations of the United States, 1951. *Undated Memorandum by the Prime Minister of Japan (Yoshida)*, FRUS, Vol. VI, pp. 833–4.
77. Schaller, *Altered States*, p. 45.
78. Ibid., p. 49

79. Ibid.
80. Ibid., p. 40.
81. Yoshida, *Yoshida Shigeru*, p. 217.
82. Schaller, *Altered States*, p. 40.
83. Ibid., p. 46.
84. Welfield, *An Empire in Eclipse*, p. 91.
85. Robert J. Watson, 1986. *History of the Joint Chiefs of Staff*, vol. 5. The JCS and National Policy, 1952–4, Washington DC, pp. 267–75.
86. Marc Gallicchio, 1991. 'The Kurils Controversy: U.S. Diplomacy in the Soviet-Japan Border Dispute, 1941–1956,' *Pacific Historical Review* 60(1): pp. 69–101.
87. Eiji Takemae, R. Ricketts Swann and J.W. Dower, 2002. *The Allied Occupation of Japan*. New York and London: Continuum.
88. Ibid.
89. Gallicchio, 'The Kurils Controversy', pp. 87–8, and p. 100.
90. Schaller, *Altered States*, p. 63.
91. Ibid., p. 64.
92. Tadashi Aruga, 1989. 'Security Treaty Revision,' in Akira Iriye and Warren I. Cohen (eds), *The United States and Japan in the Postwar World*. Lexington, KY: University Press of Kentucky, p. 62.
93. Schaller, *Altered States*, p. 114.
94. Fuji Kamiya, 'The Northern Territories: 130 Years of Japanese Talks with Czarist Russia and the Soviet Union,' in Donald S. Zagoria (ed.), *Soviet Policy in East Asia*. New Haven and London: Yale University Press, pp. 128–9.
95. Schaller, *Altered States*, p. 116.
96. Welfield, *An Empire in Eclipse*, p. 108.
97. *New York Times*, 26 August 1956.
98. Schaller, *Altered States*, p. 123.
99. Welfield, *An Empire in Eclipse*, p. 120.
100. Aruga, 'Security Treaty Revision', p. 63.
101. Welfield, *An Empire in Eclipse*, p.144.
102. Ibid., pp. 144–5.
103. Dai Sanjuyon-Kai Kokkai, Yosan Iinkai Shugiin and Dai-Jugo Giroku, 25 March 1960, quoted in Welfield, *An Empire in Eclipse*.
104. Welfield, *An Empire in Eclipse*, p. 121. In the interview with John Welfield, Kishi said, 'I think it very strange that we cannot participate in United Nations forces to defend the peace and security of a particular region.'
105. Chung Kyungmo. 'Aru Minzokushugisha No Shogai,' (The life of a Nationalist) *Sekai*, December 1975, quoted in Welfield, *An Empire in Eclipse*, p. 121.
106. Eiji Takemae, 2002. *The Allied Occupation of Japan*. London: Continuum International Publishing Group, p. 448.
107. Ibid.
108. Morris-Suzuki, *Borderline*, p. 55.
109. Takemae, *The Allied Occupation of Japan*, p. 449.
110. Chung, *Immigration and Citizenship*.
111. Joint Chiefs of Staff, 1945, Basic Initial Post-Surrender Directive to Supreme Commander for the Allied Powers for the Occupation and Control of Japan (JCS1380/15), 3 November 1945, in Government Section, SCAP, 1949, *Political Reorientation of Japan*. Washington, DC: US Government Printing Office, pp. 429–41, http://www.ndl.go.jp/constitution/e/shiryo/01/036/036tx.html#t001 (accessed 15 March 2013).

112. Tessa Morris-Suzuki, 2010. *Borderline Japan: Foreigners and Frontier Controls in the Postwar Era*. Cambridge: Cambridge University Press, p. 60.
113. Welfield, *An Empire in Eclipse*, p. 92.
114. Watt, *When Empire Comes Home*, pp. 94-5.
115. Edward D. Wagner, 1951. *The Korean Minority in Japan, 1904-1950*. New York: Institute of Pacific Relations.
116. Takemae, *The Allied Occupation of Japan*.
117. Ibid., p. 449
118. Ibid.
119. Ibid.
120. Kashiwazaki, 'The Politics of Legal Status', p. 24.
121. See Chapter 2 on Japan's colonial legacy. The Governors-General of Korea were responsible to the Emperor alone, and even after that time were never fully subordinate to either the Japanese cabinet or the prime minister. Additionally, in order to exercise complete control over all civilian and military affairs, the Governor-General was authorized to issue *Seirei* (executive ordinances) which carried the same effect as laws passed by the Imperial Diet.
122. Kashiwazaki, 'The Politics of Legal Status', p. 21.
123. Yasuaki Onuma, 1979d. *Shutsunyukoku kanri hosei no seiritsukatei (The Process of the Establishment of the Immigration Control System)*, 14, Horitsu jiho 51(5): pp. 100-6.
124. Hiromitsu Inokuchi, 2000. 'Korean ethnic schools in Occupied Japan, 1945-52,' in Sonia Ryang (eds), *Koreans in Japan: Critical Voices from the Margin*. New York: Routledge.
125. Takemae, *The Allied Occupation of Japan*, p. 462.
126. The League set itself three major objectives: to help compatriots return to their homeland, to secure social equality and livelihood rights for those in Japan and to establish an independent government in Korea. *Choren's* top leadership was closely associated with the Japan Communist Party, but its rank and file included both leftist sympathizers as well as anti-Communists. In 1946, the rightist groups split off from the *Choren*, and formed the Korean Residents' Union in Japan (*Mindan*) that became pro-South Korean government. See Takemae, *The Allied Occupation of Japan*.
127. Mark Caprio, 2008. 'The Cold War Explodes in Kobe – The 1948 Korean Ethnic School Riots and US Occupation Authorities,' *The Asia-Pacific Journal: Japan Focus* 48-2-08.
128. Sung-Hwa Cheong, 1991. *The Politics of Anti-Japanese Sentiment in Korea: Japanese-South Korean Relations under American Occupation, 1945-1952*. New York: Greenwood Press.
129. Takemae, *The Allied Occupation of Japan*, pp. 495-6.
130. Ibid., p. 497
131. Ibid.
132. Foreign Relations of the United States, Asia and the Pacific (in two parts), 1951, *Memorandum of Conversation, by Mr. Robert A. Fearey of the Office of Northeast Asian Affairs*. FRUS, Vol. VI, pp. 1007-08. 23 April 1951.
133. Takemae, *The Allied Occupation of Japan*, p. 510; see also, Samuel S. Stratton, 1948. 'The Far Eastern Commission,' *International Organization* 2(1): 1-18: 'The Far Eastern Commission, a body made up of eleven nations and charged with the responsibility of prescribing the policies, principles and standards by which Japan shall fulfill the terms of her surrender belongs to this group.'
134. Takemae, *The Allied Occupation of Japan*, p. 511.

135. Morris-Suzuki, *Borderline*; see also Chung, *Immigration and Citizenship*.
136. Morris-Suzuki, *Borderline*, p. 110.
137. Ibid. Under this system each foreigner was to have a full set of ten fingerprints taken each time, which served to provide a store of data against which police could match prints found at crime scenes. This allowed Zainichi to be treated as potential criminals.
138. Mylonas, *The Politics of Nation-Building*.
139. Ibid., p. 22.
140. Morris-Suzuki, *Borderline*, pp. 194–222.
141. Ibid.
142. Ibid.
143. *Yomiuri* Newspaper, 1959. 'Minority in Japan: Zainichi Korean who migrated for survival but 80% of them unemployment status,' 14 March 1959, https://database.yomiuri.co.jp/rekishikan/viewerMtsStart.action?objectId=sTE7HhiNcvXGMajuzPfdO2k6LDhVoXJ%2FqDRkf7zEyCg%3D (accessed 2 December 2014).
144. National Archives and Records Administrations (NARA), 1959. *Telegram from US Ambassador, MacArthur to Secretary of State, Washington*, 7 February 1959, decimal file no. 694.95B/759, parts 1 and 2, quoted in Tessa Morris-Suzuki, 2007. *Exodus to North Korea: Shadows from Japan's Cold War*. Lanham, MD: Rowman & Littlefield Publishers.
145. Ibid.
146. *Yomiuri* Newspaper, 1955. 'Not overthrowing Japanese Government: North Korean leader's announcement to Zainichi Koreans,' 23 October 1955, https://database.yomiuri.co.jp/rekishikan/viewerMtsStart.action?objectId=l0au26ZbyOEVGbsWuYL7N9K8KjfJOAjWdnw3Vrq5IQg%3D (accessed 2 December 2014).
147. Japan Red Cross Society, 1956. *The Repatriation Problem of Certain Koreans Residing in Japan*. Tokyo: Japan Red Cross Society.
148. Morris-Suzuki, *Borderline*, pp. 194–222
149. Ibid.
150. Morris-Suzuki, *Exodus to North Korea*.
151. Morris-Suzuki, *Borderline*, p. 202.
152. Ibid., p. 204.
153. Morris-Suzuki, *Exodus to North Korea*, p. 202.
154. Ibid., p. 203.
155. Morris-Suzuki, *Borderline*, p. 205.
156. Alexander Kim, 'Soviet Policy in North Korea,' p. 249.
157. Morris-Suzuki, *Borderline*, p. 205.
158. *Yomiuri* Newspaper, 1959. 'Welcoming for Repatriates in North Korea,' 8 November 1959: https://database.yomiuri.co.jp/rekishikan/viewerMtsStart.action?objectId=ZZ5F9Z8EFJ2hq03JEEZJgWk6LDhVoXJ%2FqDRkf7zEyCg%3D (accessed 2 December 2014).
159. Morris-Suzuki, *Borderline*, p. 208.
160. *Yomiuri* Newspaper, 1960. 'North Korea Repatriation,' 24 November 1960: https://database.yomiuri.co.jp/rekishikan/viewerMtsStart.action?objectId=LHu1psKETCf1SvDlxxXT3pB6RpTsc7BlXVOYzDte0E%3D (accessed 31 August 2013).
161. Morris-Suzuki, *Borderline*, p. 216.
162. Interview in 2000.
163. Morris-Suzuki, *Exodus to North Korea*, p. 238.
164. Morris-Suzuki, *Borderline*, p. 216.
165. Morris-Suzuki, *Exodus to North Korea*, pp. 206–7.
166. Michael Schaller, 1997. *Altered States: The United States and Japan Since the Occupation*. New York: Oxford University Press, p. 162.

167. Foreign Relations of the United States, 1960. *Telegram From the Embassy in Japan to the Department of State*. Foreign Relations of the United States 1958–60. Japan, Korea, Vol. XVIII, p. 422, 16 December 1960.
168. Victor D. Cha, 1999. *Alignment despite Antagonism: The United States-Korea-Japan Security Triangle*. Stanford: Stanford University Press, p. 29.
169. Lee and De Vos, *Koreans in Japan*, p. 92.
170. Ibid.
171. Ibid., p. 95.
172. National Archives and Records Administrations (NARA), 1960, *The Repatriation Problem Revisited* (Confidential, Memorandum), 13 May 1960, Japan and the US, 1960–1976, JU00040, NARA declassified NND959269, http://nsarchive.chadwyck.com/nsa/documents/JU/00040/all.pdf (accessed 15 March 2013).

Chapter 4: Alliance Cohesion Matters: Japan's Policy towards the Zainichi during the Cold War Era (1965–80s)

1. Gellner, *Nations and Nationalism*.
2. Morrow, 'Alliances and Asymmetry,' pp. 908–9.
3. Walt, 'Why Alliances Endure or Collapse,' p. 164.
4. Ibid., p. 165.
5. Morrow, 'Alliances and Asymmetry,' p. 910.
6. Kent E. Calder, 2007. 'Beneath the Eagle's Wings? The Political Economy of Northeast Asian Burden-Sharing in Comparative Perspective,' *Asian Security*, 2(3): pp. 148–73.
7. See Brubaker, *Nationalism Reframed*; and Mylonas, *The Politics of Nation-Building*.
8. Cha, *Alignment despite Antagonism*, p. 73.
9. Morrow, 'Alliances and Asymmetry,' p. 910.
10. Cho, *Korea in World Politics*, p. 716.
11. Cha, *Alignment despite Antagonism*, p. 60.
12. Okonogi, 'A Japanese Perspective on Korea-Japan Relations,' p. 22.
13. Cho, *Korea in World Politics*, p. 718.
14. Lee, *A Troubled Peace*, p. 54.
15. Ahn, 'The United States and Korean-Japanese Relations,' p. 133.
16. Sung-joo Han, 1983. 'South Korea and the United States: Past, Present, and Future,' in Gerald L. Curtis and Sung-joo Han (eds), *The U.S.-South Korean Alliance: Evolving Patterns in Security Relations*. Toronto: D.C. Heath and Company, pp. 208–17.
17. Hyogo Fujii, 1965. 'Nikkan Keizai Kyoroku heno Kozo' ['Ideas on Japan-Korean Economic Cooperation'], *Jiyu*, pp. 84–91.
18. Welfield, *An Empire in Eclipse*, p. 192.
19. Cha, *Alignment despite Antagonism*, p. 54.
20. Edwin O. Reischauer, 1967. 'Our Dialogue with Japan,' *Foreign Affairs*, pp. 215–28.
21. John K. Emmerson, 1978. *The Japanese Thread: A Life in the US Foreign Service*. New York: Holt, Rinehart and Winston.
22. The President emphasized the great progress that had taken place in East Asia this past decade or so and stressed that the East Asian countries could now take on a large share of their own defense. (See Richard Nixon, 1967. 'Asia After Vietnam,' *Foreign Affairs* October 1967 Issue, https://www.foreignaffairs.com/articles/asia/1967-10-01/asia-after-viet-nam (accessed 23 May 2015); also available from *Department of State Bulletin*, 15 December 1969).

23. Cha, *Alignment despite Antagonism*, p. 76; see also, Chong-Sik Lee, 1985. *Japan and Korea: The Political Dimension*. Stanford, CA: Hoover Institution Press, p. 71.
24. Okonogi, 'A Japanese Perspective on Korea-Japan Relations,' pp. 26-7.
25. Ibid., pp. 27-8.
26. Ryang, *North Koreans in Japan*, p. 124.
27. Ibid.
28. *Dong-A Ilbo*, 19 December 1966.
29. Tomas Hammar, 1990. *Democracy and the Nation State: Aliens, Denizens, and Citizens in a World of International Migration*. Aldershot: Avebury.
30. The National Voting rights started because of a Zainichi Korean named Li Konu, who won a judgment in 2007 after appealing to the Constitutional Court to seek voting rights for overseas Koreans in the Korean government elections.
31. *Japan Times*, 13 June 1963.
32. Kyung Moon Hwang, 2010. *A History of Korea: An Episodic Narrative*. Hampshire: Palgrave Macmillan, pp. 229-30.
33. Morris-Suzuki, *Borderline Japan*, p. 230.
34. Welfield, *An Empire in Eclipse*, p. 208.
35. Rodong Shinmun, 26 June 1965 quoted in Ryang, *North Koreans in Japan*.
36. Udai Fujishima, 1966. 'Waga Michio Yuku Jishin' ['Confidence of Going My Own Way'], *Economist*, Vol. XLIV, pp. 36-8.
37. Ryang, *North Koreans in Japan*, p. 116.
38. Ibid., p. 117.
39. Lee and De Vos, *Koreans in Japan*, pp. 146-7.
40. *Yomiuri Newspaper*, 10 November 1965, https://database.yomiuri.co.jp/rekishikan/viewerMtsStart.action?objectId=67R1If%2B%2BbFFI4yQH5DJkvKdmFjPuBLg98YSU4ORfSKM%3D (accessed 3 August 2015).
41. Ibid.
42. Chung, *Immigration and Citizenship*, p. 89.
43. *Asahi Newspaper*, 7 March 1965, http://database.asahi.com/library2e/smendb/d-image-frameset-main.php (accessed 3 August 2015).
44. Lee and De Vos, *Koreans in Japan*, p. 151.
45. Morris-Suzuki, *Borderline Japan*, p. 230; see also Ryang, *North Koreans in Japan*, p. 124. It was agreed that the sides would review these terms in 1991.
46. Stephen M. Walt, 1985. 'Formation and the Balance of World Power,' *International Security* 9(4): pp. 3-43.
47. Seung-Young Kim, 2012. 'Balancing Security Interest and the "Mission" to Promote Democracy: American Diplomacy Toward South Korea Since 1969,' in Robert A. Wampler (eds), *Trilateralism and Beyond: Great Power Politics and the Korean Security Dilemma During and After the Cold War*. Kent, OH: Kent State University Press, p. 51.
48. Cha, *Alignment despite Antagonism*, p. 101.
49. Lee Chae-Jin and Doo-Bok Park, 1996. *China and Korea: Dynamic Relations*. Stanford, CA: Hoover Institution Press, p. 65.
50. William Burr, 1998. *The Kissinger Transcripts*. New York: The New Press, p. 29.
51. Ibid., p. 32; see also Cha, *Alignment despite Antagonism*, p. 102.
52. Walt, 'Formation and the Balance,' p. 6.
53. Burr, *The Kissinger Transcripts*, p. 29.
54. Ibid., pp. 29, 40.
55. Ibid., p. 38.
56. Snyder, 'Alliances, Balance, and Stability,' p. 128; see also Morrow, 'Alliances and Asymmetry,' for a discussion of the trade-offs between security and autonomy, pp. 926-7.

57. Snyder, 'Alliances, Balance, and Stability,' p. 125.
58. Burr, *The Kissinger Transcripts*, p. 43.
59. Ibid., p. 33.
60. Cha, *Alignment despite Antagonism*, p. 108; see Lee and Park, *China and Korea*, p. 66. According to Kissinger's report, Mao and Zhou specifically disavowed China's threatening Japan or South Korea and refrained from asking Nixon to withdraw US forces from South Korea.
61. Welfield, *An Empire in Eclipse*, p. 300.
62. Ibid., p. 314.
63. Seung-Young Kim, 'Miki Takeo's Initiative,' p. 385.
64. Ibid., p. 384.
65. Lee and Park, *China and Korea*, p. 106.
66. National Archives, Nixon Presidential Materials, NSC Files, Box 364, Subject Files, National Security Decision Memoranda, Nos. 97–144, 1970. Telegram From the Embassy in Korea to the Department of State. FRUS 1969–1976. Volume XIX, Part 1, Korea, 1969–1972, 15 June 1970, http://history.state.gov/historicaldocuments/frus1969-76v19p1/d61 (accessed 2 December 2014).
67. National Archives, Nixon Presidential Materials, NSC Files, Box 757, Presidential Correspondence 1969–1974, Korea, President Chung Hee Park, 1971. *Letter From President Nixon to Korean President Park*. FRUS 1969–1976. Volume XIX, Part 1, Korea, 1969–1972, 16 July 1971, http://history.state.gov/historicaldocuments/frus1969-76v19p1/d100 (accessed 2 December 2014).
68. Mark L. Clifford, 1998. *Troubled Tiger*. New York: An East Gate Book, p. 76.
69. Ibid., pp. 76–98.
70. Lee, *A Troubled Peace*, p. 75.
71. Ibid.
72. Cha, *Alignment despite Antagonism*, p. 121.
73. Lee, *Japan and Korea*, p. 79.
74. *Yomiuri Newspaper*, 1972. 'Tax-exempt status applicable, for Chosen Soren,' 2 July 1972, https://database.yomiuri.co.jp/rekishikan/viewerMtsStart.action?objectId=hESgNt0Pm%2FwJiwph1UJ2VHhu%2BboWjronX%2FW6d9Lc9vU%3D (accessed 2 December 2014).
75. *Japan Times*, 18 August 1972.
76. *Japan Times*, 17 September 1971.
77. Apichai Shipper, 2009. 'Nationalisms of and Against Zainichi Koreans in Japan,' *Asian Politics & Policy* 2(1): pp. 55–75, p. 61.
78. *Korea Herald*, 29 August 1974.
79. Cha, *Alignment despite Antagonism*, p. 130.
80. Morris-Suzuki, *Borderline Japan*, p. 231.
81. Sakanaka Hidenori, 1989. *Kongo no Shutsunyukoku Kanri no Arikata ni tsuite*. Tokyo: Kodansha.
82. Morris-Suzuki, *Borderline Japan* p. 232.
83. Benedict Anderson, 1994. 'Exodus,' *Critical Inquiry* 20(2): pp. 314–27. See also Anderson, *Imagined Communities*.
84. Sang-Jung Kang, 2011. 'Tunneling Through Nationalism: The Phenomenology of a Certain Nationalist,' *The Asia-Pacific Journal* 9, 36(2).
85. Brubaker, 'The "Diaspora" Diaspora,' p. 12.
86. Kang, 'Tunnelling Through Nationalism'.
87. Brubaker, 'The "Diaspora" Diaspora'.

NOTES TO PAGES 136–145

88. *Yomiuri Newspaper*, 1974. 'Prison note of Mun Se Kwang,' 14 December 1974, https://database.yomiuri.co.jp/rekishikan/viewerMtsStart.action?objectId=O9Ljd%2FAOrNrhWaTpJ6Dm9ixuYycY5UrOe9PhRI1SEwc%3D (accessed 2 December 2014).
89. Ryang, *North Koreans in Japan*, p. 107.
90. Anthony Giddens, 1979. *Central Problems in Social Theory: Action, Structure and Contradiction in Social Analysis*. London: Macmillan.
91. Ryang, *North Koreans in Japan*, p. 102.
92. Cha, *Alignment despite Antagonism*, p. 123.
93. Ibid.
94. Burr, *The Kissinger Transcripts*, p. 288.
95. Walt, 'Alliance Formation,' p. 40.
96. Calder, 'Beneath the Eagle's Wings?,' p. 149.
97. Stephen D. Krasner, 1986. 'Trade Conflicts and the Common Defence: The United States and Japan,' *Political Science Quarterly* 101(5): pp. 787–806. See also Robert O. Keohane, 1978. 'American Policy and Trade-Growth Struggle,' *International Security* 3(3): pp. 20–43.
98. Calder, 'Beneath the Eagle's Wings?,' p. 149.
99. Cha, *Alignment despite Antagonism*, p. 154.
100. Thomas R. H. Havens, 1987. *Fire Across the Sea: The Vietnam War and Japan, 1965–1975*. Princeton, NJ: Princeton University Press.
101. Lee, *A Troubled Peace*, pp. 81–5.
102. Ibid., p. 85.
103. Chong-Sik Lee, 1985. *Japan and Korea: The Political Dimension*. Stanford: Hoover Institution Press, p. 94. See also Cha, *Alignment despite Antagonism*, p. 154.
104. Cha, *Alignment despite Antagonism*, p. 157.
105. *Asahi Newspaper*, 17 September 1975, p. 5, http://database.asahi.com/library2e/smendb/d-image-frameset-main.php (accessed 2 December 2014).
106. Lee and Park, *China and Korea*, p. 69.
107. Lee, *A Troubled Peace*, p. 90.
108. Lee, *Japan and Korea*, p. 99.
109. Lee, *A Troubled Peace*, p. 94.
110. Krasner, 'Trade Conflicts'; and see also Keohane, 'American Policy and Trade-Growth,' pp. 20–43.
111. Hahei Chekku Henshū Iinkai, 1997. *Kore Ga Beigun e No 'Omoiyari Yosan' Da!: 'Nichi-Bei Anpo'-dokuhon*. Shohan; Tōkyō: Shakai Hyōronsha, p. 68.
112. Calder, 'Beneath the Eagle's Wings?,' pp. 164–6.
113. *Yomiuri Newspaper*, 1975, 'We Strongly Support Revolution in South,' 19 April 1975, http://database.asahi.com/library2e/smendb/d-image-frameset-main.php (accessed 2 December 2014).
114. *Asahi Newspaper*, 6 September 1974, p. 1, http://database.asahi.com/library2e/smendb/d-image-frameset-main.php (accessed 2 December 2014).
115. Cha, *Alignment despite Antagonism*, p. 162; see also Lee, *Japan and Jorea*, p. 95; and *Tong-a Ilbo*, 29 May 1975.
116. *Asahi Newspaper*, 10 November 1977, p. 2, http://database.asahi.com/library2e/smendb/d-image-frameset-main.php (accessed 2 December 2014).
117. Cha, *Alignment despite Antagonism*, p. 123.
118. *Asahi Newspaper*, 12 October 1973, p. 3, http://database.asahi.com/library2e/smendb/d-image-frameset-main.php (accessed 2 December 2014).
119. Clifford, *Troubled Tiger*, pp. 84–5.

120. *Asahi Newspaper*, 4 July 1977, p. 1, http://database.asahi.com/library2e/smendb/d-image-frameset-main.php (accessed 2 December 2014).
121. *Asahi Newspaper*, 22 November 1975, p. 11, http://database.asahi.com/library2e/smendb/d-image-frameset-main.php (accessed 2 December 2014).
122. Eika Tai, 2009. 'Between Assimilation and Transnationalism: The Debate on Nationality Acquisition among Koreans in Japan,' *Social Identities* 15(5): pp. 609–29.
123. Sang-Jung Kang, 1991. '*Hoho toshiteno zainichi* [Zainichi as method – in response to Mr. Yang Tae-Ho's critique]'. Osaka: Kaifusha.
124. Tai, 'Between Assimilation and Transnationalism,' p. 616.
125. Morris-Suzuki, *Borderline Japan*, p. 232.
126. Tai, 'Between Assimilation and Transnationalism,' p. 614.
127. Il Park, 1999. *Zainichi to iu ikikata (A Way of life as Zainichi)*. Tokyo: Kodansha.
128. Chung, *Immigration and Citizenship*, pp. 96–100.
129. Ibid.
130. Walt, 'Alliance Formation,' p. 7: 'John F. Kennedy claimed that if the United States were to falter, the whole world … would inevitably begin to move toward the Communist bloc.' Quoted from Seyom Brown, 1968. *The Faces of Power*. New York: Columbia University Press.
131. Caspar Weinberger, 1986. 'US Defence Strategy,' *Foreign Affairs* 64(4): pp. 675–97.
132. Lee, *A Troubled Peace*, p. 113.
133. George P. Shultz, 1993. *Turmoil and Triumph: My Years as Secretary of State*. New York: Charles Scribner's Sons, p. 975.
134. Cha, *Alignment despite Antagonism*, p. 172.
135. The Digital National Security Archive (DNSA) Online, 1983. *Defence Estimative Brief, Prospects for South Korea*. 28 March 1983, NLS F95-033/1 #180, DEB-29-83, http://nsarchive.chadwyck.com/quick/displayMultiItemImages.do?Multi=yes&ResultsID=14CDC24B77B&queryType=quick&QueryName=cat&ItemID=CKO00438&ItemNumber=924.
136. Alexander Haig's memorandum for the president, 1981. *Your Meeting with Chun Doo-Hwan, President of the Republic of Korea*, 29 January 1981. Declassified in 1993, quoted in Lee, *A Troubled Peace*, p. 113.
137. Morrow, 'Alliances and Asymmetry,' p. 913.
138. Lee, *A Troubled Peace*, p. 114.
139. Okonogi, 'A Japanese Perspective,' p. 31.
140. *Sekai Shunho*, 2 June 1981, quoted in Okonogi, 'A Japanese Perspective,' p. 32.
141. Lee, *Japan and Korea*, p. 119.
142. Ahn, 'The United States and Korean-Japanese Relations,' pp. 148–9.
143. Cha, *Alignment despite Antagonism*, p. 185.
144. *Yomiuri Newspaper*, 22 August 1981, p. 1, https://database.yomiuri.co.jp/rekishikan/viewerMtsStart.action?objectId=NaxrGtBsalz0s01PLdgkCnpB6RpTsc7BIXVOYzDte0E%3D (accessed 2 December 2014).
145. Clifford, *Troubled Tiger*, p. 163.
146. Bruce Cumings, 1984. 'The Legacy of Japanese Colonialism in Korea,' in Ramon H. Myers and Mark R. Peattie (eds), *The Japanese Colonial Empire, 1895–1945*. Princeton: Princeton University Press, 1984; see also Hong Nak Kim 1982. 'After the Park Assassination,' *Journal of Northeast Asian Studies* 1(4): pp. 71–90.
147. Cumings, 'The Legacy of Japanese Colonialism'.
148. Cha, *Alignment despite Antagonism*, p. 186.
149. Donald N. Clark, 1980. *The Kwangju Regime in South Korea*. Boulder: Westview Press.
150. Lee, *Japan and Korea*, p. 143.

151. Ibid., pp. 145–7.
152. Cha, *Alignment despite Antagonism*, p. 188.
153. Paul F. Gardner, 1999. *The Nakasone Government: Prospects and Problems*. United States Department of State Review, Bureau of Intelligence and Research: Issues Paper, Date/Case ID: 8 September 1999, 199501403, http://nsarchive.chadwyck.com/nsa/documents/JA/01059/all.pdf (accessed 22 March 2014).
154. Akio Watanabe, 1985. 'Political Change in Japan and Korea-Japan Relations,' in *Korea and Japan in World Politics*. Seoul: The Korean Association of International Relations.
155. Cha, *Alignment despite Antagonism*, p. 177.
156. Gardner, *The Nakasone Government*.
157. Robert H. Miller, 2001. *Nakasone's Visit to Washington: The View from Japan*. United States Department of State Review Authority: Date/Case ID: 5 June 2001, 199504020, http://nsarchive.chadwyck.com/nsa/documents/JA/01400/all.pdf (accessed 22 March 2014).
158. Gardner, *The Nakasone Government*.
159. Miller, *Nakasone's Visit to Washington*.
160. Central Intelligence Agency, 1984. 'Briefing Materials for the President's Meeting with Prime Minister Nakasone, 2 January 1985,' 21 December 1984, Washington, DC, http://www.foia.cia.gov/sites/default/files/document_conversions/89801/DOC_0000309669.pdf (accessed 24 March 2014).
161. *Asahi Newspaper*, 9 September, 1984, p. 1, http://database.asahi.com/library2e/smendb/d-image-frameset-main.php (accessed 24 March 2014).
162. Ibid.
163. Evelyn Colbert, 1986. 'Japan and Republic of Korea,' *Asian Survey* 26(3): pp. 273–91, p. 279.
164. *Asahi Newspaper*, 31 October 1984, p. 1, http://database.asahi.com/library2e/smendb/d-image-frameset-main.php (accessed 14 March 2014). The bombing in Rangoon in October 1983 killed 17 prominent South Koreans. When the official Burmese government investigation revealed beyond any doubt that North Korea was responsible for the bomb attack on the South Korean delegation in the Burmese capital, then Nakasone government decided to take a stern stance in its relations with North Korea. See Hong N. Kim, 1987. 'Japanese-Korean Relations in the 1980s,' *Asian Survey* 27(5): pp. 497–514.
165. Central Intelligence Agency, 1984. 'Briefing Materials for the President's Meeting with Prime Minister Nakasone, 2 January 1985,' 21 December 1984, Washington, DC, http://www.foia.cia.gov/sites/default/files/document_conversions/89801/DOC_0000309669.pdf (accessed 24 March 2014).
166. Ryang, *North Koreans in Japan*.
167. E. Tai, 2004. 'Koriakei Nihonjin to iu sentaku (A Choice to be Korean Japanese),' *Chuokoron*, pp. 360–2.
168. Rogers Brubaker, 2001. 'The Return of Assimilation? Changing Perspectives on Immigration and its Sequels in France, Germany, and the United States,' *Ethnic and Racial Studies* 24(4): pp. 531–48, p. 538.
169. Brubaker, 'The Return of Assimilation?', pp. 533–4.
170. E. Tai, 2009. 'Between Assimilation and Transnationalism: The Debate on Nationality Acquisition among Koreans in Japan,' *Social Identities* 15(5): pp. 609–29, p. 616.
171. Rynag, *North Koreans in Japan*, p. 125.
172. Ryang, 'Introduction,' p. 67.
173. Ibid.
174. Rynag, *North Koreans in Japan*, p. 167.
175. Ibid., p. 172.
176. Central Intelligence Agency, 1984. 'Briefing Materials for the President's Meeting with Prime Minister Nakasone, 2 January 1985,' 21 December 1984, Washington,

DC, http://www.foia.cia.gov/sites/default/files/document_conversions/89801/DOC_0000309669.pdf (accessed 24 March 2014), p. 24.
177. Ibid.
178. Cha, *Alignment despite Antagonism*, p. 191.
179. Ibid., p. 193.
180. Ibid., p. 190.
181. Ibid., p. 196.
182. Ibid.
183. Ibid., p. 190.
184. Hiroki Manabe, 2013. 'INTERVIEW/ Carol Gluck: Change in Japan is a long-distance run,' *Asahi Newspaper*, 17 September 2013, http://ajw.asahi.com/article/views/opinion/AJ201309170004 (accessed 14 March 2014).
185. Cha, *Alignment despite Antagonism*, p. 56.

Chapter 5: Does Alliance Cohesion Still Matter in the New Post-Cold War (1990–2014)?

1. Brett Ashley Leeds and Burcu Savun, 2007. 'Terminating Alliances: Why do States Abrogate Agreements?,' *Journal of Politics* 69(4): pp. 1118–32.
2. Cha, *Alignment despite Antagonism*.
3. Scott Snyder, 2011. 'Korea and the US-Japan Alliance: An American Perspective,' in Takahashi Inoguchi, G. John Ikenberry and Yochiro Sato (eds), *The US-Japan Security Alliance, Regional Multilateralism*. New York: Palgrave Macmillan, p. 128.
4. Yasuyo Sakata, 2011. 'Korea and the Japan-US Alliance: A Japanese Perspective,' *The US-Japan Security Alliance*, p. 95.
5. Marie Soderberg, 2011. 'Japan-South Korea Relations at a Crossroads,' in Marie Soderberg (ed.), *Changing Power Relations in Northeast Asia*. London: Routledge.
6. Sakata, 'Korea and the Japan-US Alliance: A Japanese Perspective,' p. 95.
7. Nicholas Eberstadt, Aaron L. Friedberg and Geun Lee (eds), 2008. 'Introduction: What If? A World Without the US-ROK Alliance,' *Asia Policy* (5): pp. 2–5.
8. Sakata, 'Korea and the Japan-US Alliance: A Japanese Perspective,' p. 95.
9. Ibid., p. 94.
10. Michel Auslin and Christopher Griffin, 2008. *Securing Freedom: The US-Japanese Alliance in a New Era: A Report of the American Enterprise Institute*. Washington DC: American Enterprise Institute, p. 14.
11. Sakata, 'Korea and the Japan-US Alliance: A Japanese Perspective,' p. 108.
12. Tomohito Shinoda, 2011. 'Costs and Benefits: US-Japan Alliance,' in *The US-Japan Security Alliance*, p. 20.
13. Ibid., p. 21.
14. Snyder, 'Korea and the Japan-US Alliance: An American Perspective,' p. 119.
15. Morrow, 'Alliances and Asymmetry,' pp. 913–16.
16. Leeds and Savun, 'Terminating Alliances,' p. 1120.
17. Ralph A. Cossa, 1999. 'Preface,' in Ralph A. Cossa (eds), *US-Korea-Japan Relations: Building Toward a Virtual Alliance*, Washington, DC: The CSIS Press, p. 22.
18. Michael Jonathan Green, 2001. *Japan's Reluctant Realism*. New York: Palgrave, p. 12.
19. Sakata, 'Korea and the Japan-US Alliance: A Japanese Perspective,' p. 92.
20. Noboru Yamaguchi, 1999. 'Trilateral Security Cooperation: Opportunities, Challenges, and Tasks,' in Ralph A. Cossa (ed.), *US-Korea-Japan Relations: Building Toward a Virtual Alliance*. Washington, DC: The CSIS Press, p. 11.

21. Green, *Japan's Reluctant Realism*, p. 20.
22. Hisayoshi Ina, 1996. 'Fear of US-China Ties Rests on Flawed Premises,' *Nikkei Weekly*, 18 December 1996, quoted by Green, *Japan's Reluctant Realism*, p. 23.
23. Gaubatz Kurt Taylor, 1996. 'Democratic States and Commitment in International Relations,' *International Organization* (1): pp. 109–39.
24. Leeds and Savun, 'Terminating Alliances,' p. 1122.
25. Snyder, 'Korea and the Japan-US Alliance: An American Perspective,' p. 132.
26. Christopher Griffin and Michael Auslin, 2008. 'Time for Trilateralism?', *Asian Outlook* (2). Washington, DC: American Enterprise Institute for Public Policy Research.
27. Ralph A. Cossa, 1999. 'Preface,' in *US-Korea-Japan Relations*, p. 16.
28. Green, *Japan's Reluctant Realism*, p. 134.
29. There are reports of thousands of women who were allegedly forcibly recruited into the prostitution (sexual slaves) during World War II. The number of these women is reputed to be several hundred thousand. The estimate of 200,000 was acknowledged in a 1993 Japanese government report. Most, perhaps 80 per cent, were Korean although they came from all over the Pacific, including several Dutch women. See Jane W. Yamazaki, 2006. *Japanese Apologies for World War II*. London: Routledge, p. 29.
30. Yoshimi Yoshiaki, 2001. 'State Crime Should Be Compensated by the State,' *Asahi Shinbun*, 22 August 1995, quoted by Green, *Japan's Reluctant Realism*, p. 134.
31. Green, *Japan's Reluctant Realism*, p. 134.
32. Jennifer Lind, 2008. *Sorry States*. Ithaca; London: Cornell University Press.
33. Manabe, 'INTERVIEW/ Carol Gluck' *Asahi Newspaper*. See also, Carol Gluck, 2007. 'Operations of Memory,' in Sheila Miyoshi Jager and Rana Mitter (eds), *Ruptured Histories: War, Memory and the Post-Cold War in Asia*. London: Harvard University Press.
34. Green, *Japan's Reluctant Realism*, p. 136.
35. Takashi Uemura, 1998. 'Open on Japanese Culture,' *Asahi Newspaper*, 20 October 1998, p. 2, http://database.asahi.com/library2e/main/start.php (accessed 14 March 2014).
36. Green, *Japan's Reluctant Realism*, p. 138.
37. Sakata, 'Korea and the Japan-US Alliance: A Japanese Perspective,' p. 99.
38. Leeds and Savun, 'Terminating Alliances,' p. 1120.
39. Morrow, 'Alliances, Credibility, and Peacetime Costs,' p. 272.
40. Yoichi Funabashi, 1999. *Alliance Adrift*. Washington, DC: Council on Foreign Relations.
41. Green, *Japan's Reluctant Realism*, p. 118.
42. Ibid., p. 119.
43. Michitaka Narushige, 2009. *North Korea's Military and Diplomatic Campaigns, 1966–2008*. New York: Routledge.
44. Sakata, 'Korea and the Japan-US Alliance: A Japanese Perspective,' p. 97.
45. Ibid., p. 98.
46. Green, *Japan's Reluctant Realism*, p. 123.
47. Ibid., p. 127.
48. Ministry of Foreign Affairs of Japan, 1998. *Japan-Republic of Korea Joint Declaration: A New Japan-Republic of Korea Partnership Toward the Twenty-First Century*, 8 October 1998, http://www.mofa.go.jp/region/asia-paci/korea/joint9810.html (accessed 15 March 2014).
49. Bumsoo Kim, 2006. 'From Exclusion to Inclusion? The Legal Treatment of "Foreigners" in Contemporary Japan,' *Immigrants & Minorities* (1): pp. 51–73.
50. Ibid., p. 66.
51. Ryang, *North Koreans in Japan*, p. 125.

52. Tomohito Shinoda, 2011. 'Costs and Benefits: US-Japan Alliance,' in Takahashi Inoguchi, G. John Ikenberry and Yochiro Sato (eds), *The US-Japan Security Alliance, Regional Multilateralism*. New York: Palgrave Macmillan, p. 20.
53. In 1990, South Korea-Soviet Union Normalization Treaty, in 1992, China-South Korea Normalization Treaty were enacted.
54. Bumsoo Kim, 'From Exclusion to Inclusion?,' p. 66.
55. Ryang, *North Koreans in Japan*, p. 125.
56. Pachinko is a mechanical game originating in Japan and is used as a recreational arcade game and much more frequently as a gambling device, filling a Japanese gambling niche comparable to that of the slot machine in Western gaming.
57. Bumsoo Kim, 2011. 'Changes in the Socio-Economic Position of Zainichi Koreans: A Historical Overview,' *Social Science Japan Journal* 14(2): pp. 233-45, p. 6.
58. Ibid.
59. Nicholas Eberstadt, 1996. 'Financial Transfers from Japan to North Korea: Estimating the Unreported Flows,' *Asian Survey* 36(5): pp. 523-42, p. 523.
60. Green, *Japan's Reluctant Realism*, p. 117.
61. Ibid.
62. Ibid., p. 119.
63. By 1992 Japanese police investigations of missing persons along the Sea of Japan coastline pointed that this provocation might be North Korean's intelligence' conduct. See Green, *Japan's Reluctant Realism*, p. 119.
64. Don Oberdorfer, 2001. *Two Koreas*. New York: Basic Books, p. 453.
65. Sakata, 'Korea and the Japan-US Alliance: A Japanese Perspective,' p. 95.
66. Green, *Japan's Reluctant Realism*, p. 119.
67. Ibid., p. 123.
68. Sakata, 'Korea and the Japan-US Alliance: A Japanese Perspective,' p. 100.
69. KEDO Press, 1998. 'KEDO Executive Board Agrees on Cost-Sharing for Light Water Reactor Project,' 10 November 1998, quoted in Green, *Japan's Reluctant Realism*, p. 125; see also Global Reporting Network Publications 1998. 'Politics and the Agreed Framework: North Korean Deal On Thin Ice?,' *Global Beat Issue Brief*, No. 44, 10 November 1998, http://www.bu.edu/globalbeat/pubs/ib44.html (accessed 24 May 2015).
70. Green, *Japan's Reluctant Realism*, p. 124.
71. Ingyu Oh, 2011. 'New Politico-Cultural Discourse in East Asia?,' in Marie Soderberg (ed.), *Changing Power Relations in Northeast Asia*. London: Routledge.
72. Tai, 'Between Assimilation and Transnationalism,' p. 618.
73. Kashiwazaki, 'The Foreigner Category,' p. 145.
74. Takeyuki Tsuda, 2008. 'Local Citizenship and Foreign Workers in Japan,' *Japan Focus*.
75. Jung-Sun Park and Paul Y. Chang, 2005. 'Contention in the Construction of a Global Korean Community: The Case of the Overseas Korean Act,' *The Journal of Korean Studies* 10(1): pp. 1-27.
76. Y. Kou, 2001. 'Nihon Kokuseki todokede hoan to zainichi korian no sentaku (The Proposal for an Automatic Access to Japanese Nationality and Resident Koreans' Choices)', *Sekai*, pp. 169-76.
77. Engin F. Istin and Patricia K. Wood, 1999. 'Redistribution, Recognition, Representation,' in Engin F. Istin and Patricia K. Wood (eds), *Citizenship and Identity*. London: Sage Publications, p. 20.
78. Sonia Ryang, 2009. 'Visible and Vulnerable,' in Ryang Sonia and Lie John (ed.), *Diaspora without Homeland: Being Korean in Japan*. London: University of California Press, p. 73.

79. Tai, 'Between Assimilation and Transnationalism,' p. 617.
80. Scott Snyder, 2009. *China's Rise and the Two Koreas*. London: Lynne Rienner Publishers, p. 5.
81. Chae-ho Chung, 2007. *Between Ally and Partner: Korea-China Relations and the United States*. New York: Columbia University Press, p. 78.
82. Ibid., p. 89.
83. Yoichi Funabashi, 2007. *The Peninsula Question: A Chronicle of the Second Nuclear Crisis*. Washington, DC: Brookings Institution, p. 8.
84. Snyder, *China's Rise and the Two Koreas*, p. 184.
85. Ibid.
86. Chung, *Between Ally and Partner*, p. 91.
87. Ibid., p. 85.
88. Gerald Segal, 1991. 'Northeast Asia: Common Security or a la Carte?,' *International Affairs* 67(4): pp. 755–67, p. 765.
89. Luye Li, 1991. 'The Current Situation in Northeast Asia: A Chinese View,' *Journal of Northeast Asian Studies* 10(1): pp. 78–81.
90. Snyder, *China's Rise and the Two Koreas*, pp. 192–3.
91. Ibid., pp. 178–80.
92. Ministry of Foreign Affairs of Japan, 2002. *Prime Minister Junichiro Koizumi's Visit to North Korea*, 17 September 2002, http://www.mofa.go.jp/region/asia-paci/n_korea/pmv0209/ (6 June accessed 2015).
93. Sakata, 'Korea and the Japan-US Alliance: A Japanese Perspective,' p. 101.
94. Funabashi, *The Peninsula Question*.
95. Ibid.
96. Sakata, 'Korea and the Japan-US Alliance: A Japanese Perspective,' p. 102.
97. Now North Korea is supposed to have tested a hydrogen bomb in 2016. Kamila Kingstone, 2016. 'North Korea's UK Ambassador: 'We want peace, but we've been victimised,' *The Guardian*, 13 January 2016.
98. Funabashi, *The Peninsula Question*, p. 426.
99. Ibid., p. 425.
100. CCTV News, 2014. 'Diplomatic talks between S. Korea, Japan end fruitless,' 3 December 2014, http://english.cntv.cn/20140312/105470.shtml.
101. The White House Office of the Press Secretary, 2014. 'Obama, Republic of Korea's President, Japanese PM in The Hague,' 25 March 2014, http://iipdigital.usembassy.gov/st/english/texttrans/2014/03/20140325296883.html?CP.rss=true#ixzz35Hgi1wsJhttp://iipdigital.usembassy.gov/st/english/texttrans/2014/03/20140325296883.html?CP.rss=true#axzz35HbvTloD.
102. Morrow, 'Alliances, Credibility, and Peacetime Costs,' p. 272.
103. Leeds and Savun, 'Terminating Alliances,' p. 1119.
104. Ibid., p. 1120.
105. Morrow, 'Alliances, Credibility, and Peacetime Costs,' p. 273.
106. *Asahi Newspaper*, 20 March 2003, p. 4, http://database.asahi.com/library2e/main/start.php (accessed 15 March 2014).
107. Auslin and Griffin, *Securing Freedom*, p. 22.
108. Michael Finnegan, 2009. *Managing Unmet Expectations in the US-Japan Alliance*. The National Bureau Asian Research: NBR Special Report #17.
109. Office of the Secretary of Defense, 2008. *Annual Report to Congress: Military Power of the People's Republic of China*, www.defenselink.mil/pubs/pdfs/China_Military_Report_08.pdf (accessed 9 July 2014).

110. Auslin and Griffin, *Securing Freedom*, p. 24.
111. Roger Cliff, Mark Burles, Michael S. Chase, Derek Eaton and Kevin L. Pollpeter, 2007. *Entering the Dragon's Lair: Chinese Antiaccess Strategies and their Implications for the United States*. Santa Monica, CA: Rand Corporation.
112. Mark Valencia, 2014. 'Asian Threats, Provocations Giving Rise to Whiffs of War,' *Japan Times*, 9 June 2014.
113. 'On September 7th, 2010 a Chinese fishing craft collided with two Japanese coastguard patrol boats near the oil-rich, uninhabited islands in the East China Sea known as Senkaku in Japan and Diaoyu, meaning "fishing platform", in China. Following the collision, coastguards boarded the trawler and arrested its crew and captain Zhan Qixiong who, as subsequent video footage revealed, had rammed his boat into the coastguard vessels.' See Joyman Lee, 2011. 'Islands of Conflict,' *History Today* 61(5): pp. 24–6.
114. *Japan Times*, 2014. 'Tokyo Hanoi Unite on Defense Ties,' 2 June 2014.
115. The annual Shangri-La Dialogue in Singapore is supposed to be the premier regional forum for 'building confidence and fostering practical security cooperation' as well as 'engendering a sense of community.'
116. Valencia, 2014. *Japan Times*, 9 June 2014.
117. *Asahi Newspaper*, 18 December 2013, p. 1, http://database.asahi.com/library2e/main/start.php (accessed 15 March 2014).
118. Ibid.
119. Ibid.
120. *Asahi Newspaper*, 17 December 2013, p. 2, http://database.asahi.com/library2e/main/start.php (accessed 15 March 2014).
121. *Japan Times*, 2 June 2014.
122. Ibid.
123. Leif-Eric Easley, 2008. 'Securing Japan's Positive Role in North-South Reconciliation: The Need for a Strong US-ROK Alliance to Reassure Japan,' *Academic Paper Series on Korea* 1, pp. 171–4.
124. Auslin and Griffin, *Securing Freedom*, p. 29.
125. Green, *Japan's Reluctant Realism*, p. 121.
126. Sakata, 'Korea and the Japan-US Alliance: A Japanese Perspective,' p. 94.
127. Ibid.
128. Cha, *Alignment despite Antagonism*; see also Green, *Japan's Reluctant Realism*.
129. Snyder, *China's Rise and the Two Koreas*; and Sakata, 'Korea and the Japan-US Alliance: A Japanese Perspective'.
130. Takashi Terada, 2011. 'Australia-Japan Security Partnership,' in Takahashi Inoguchi, G. John Ikenberry and Yochiro Sato (eds), *The US-Japan Security Alliance, Regional Multilateralism*. New York: Palgrave Macmillan, p. 222.
131. Speech by Prime Minister of Japan, Junichiro Koizumi, 'Japan and ASEAN in East Asia: A Sincere and Open Partnership'. This speech was delivered in Singapore, 14 January 2002, http://www.mofa.go.jp/region/asia-paci/pmv0201/speech.html (accessed 15 March 2014).
132. Kosuke Takahashi, 2014. 'Shinzo Abe's Nationalist Strategy With his Overt Nationalism and His Historical Rrevisionism, Shinzo Abe has a Plan for Japan,' *The Diplomat*, 13 February 2014.
133. Ibid.
134. Ibid.
135. North Korea admitted in October 2002 that it had secretly resumed a nuclear weapons programme.
136. Shipper, 'Nationalism of and against Zainichi Koreans,' p. 61.
137. Chōsen Sō ren. 2005, *Chōsen Sōren*. Tokyo: Chōsen Sōren. p. 57

138. Ibid., p. 55
139. Shipper, 'Nationalism of and against Zainichi Koreans,' p. 62.
140. Rynag, *Diaspora Without Homeland*, p. 80.
141. Ibid.
142. *Japan Times*, 31 January 2003 cited in Shipper, 'Nationalism of and against Zainichi Koreans,' p. 66.
143. Shipper, 'Nationalism of and against Zainichi Koreans,' p. 66.
144. Ibid.
145. Victor Cha, 2003. 'Japan-Korea Relations: Contemplating Sanctions,' *Comparative Connections* 5(1).
146. Shipper, 'Nationalism of and against Zainichi Koreans,' p. 63.
147. Ibid.
148. Jun Hongo, 2007. 'Court Rules Chongryun Property not Tax-Exempt,' *Japan Times*, 21 July 2007, http://www.japantimes.co.jp/news/2007/07/21/national/court-rules-chongryun-property-not-tax-exempt/#.VWcujbFwaM8 (accessed 15 March 2014).
149. Shipper, 'Nationalism of and against Zainichi Koreans,' p. 67.
150. Ibid.
151. Ibid., p. 66.
152. Ibid.
153. Ibid.
154. Ibid.
155. Ibid., p. 65.
156. Mark Schilling, 2012. 'Kazoku no Kuni (Our Homeland),' *Japan Times*, 27 July 2012.
157. Claire Lee, 'A Director's Divided Self,' *Wall Street Journal*, 14 March 2013, http://blogs.wsj.com/scene/2013/03/14/a-directors-divided-self/ (accessed 1 Oct 2016).
158. Interviewing in 2005.
159. Nicholas Eberstadt, Aaron L. Friedberg and Geun Lee (eds), 2008. 'Introduction: What If? A World Without the US-ROK Alliance,' *Asia Policy* 5, p. 2.
160. BBC World News, 2014. 'N Korea Fires Short-Range Missiles,' 9 July 2014 http://www.bbc.co.uk/news/world-asia-28223183 (accessed 9 July 2014).
161. DW AKADEMIE, 2014. 'Chinese President Xi visits Seoul in apparent snub to North Korea,' 3 July 2014, http://www.dw.de/chinese-president-xi-visits-seoul-in-apparent-snub-to-north-korea/a-17754872 (accessed 9 July 2014).
162. Lucy Williamson, 2014. 'Why is China's Leader Visiting Seoul?,' *BBC World News*, 3 July 2014, http://www.bbc.co.uk/news/world-asia-28140036 (accessed 9 July 2014).
163. John Swenson-Wright, 2014. 'What Japan's Military Shift Means,' *BBC World News*, 2 July 2014, http://www.bbc.co.uk/news/world-asia-28122791 (accessed 9 July 2014).
164. Ibid.
165. Stephen Walt, 1997. 'Why Alliances Endure or Collapse,' *Survival* 39(1): p. 910.
166. Arjun Appaudurai, 1996. *Modernity at Large: Cultural Dimensions of Globalization*. Minneapolis: University of Minnesota Press.
167. H.K. Bhabha, 1990. 'The Third Space: Interview with Homi Bhabha,' in J. Rutherford (ed.), *Identity, Community, Culture, Difference*. London: Lawrence&Wishart.
168. Sang-jung Kang, 2004. *Zainichi*. Tokyo: Kodansha.

Conclusion

1. C. Fred Bergsten, 2003. 'The Korean Diaspora and Globalization: Pat Contributions and Future Opportunities,' in C. Fred Bergsten and Inbom Choi (eds), *The Korean Diaspora in the World Economy*. Washington, DC: Institute for International Economies, p. 17.

2. See, for instance, David Chapman, 2008. *Zainichi Korean Identity and Ethnicity*. London: Routledge; Chikako Kashiwazaki, 2000. 'The Politics of Legal Status,' in Sonia Ryang (ed.) *Koreans in Japan*. London: Routledge; Fukuoka Yasunori, 1993. *Zainichi Kankoku-Chosenjin*. Tokyo: Chuo Koronsha; and Sonia Ryang, 1997. *North Koreans in Japan*. Boulder, CO: Westview Press.
3. Mylonas, *The Politics of Nation-Building*.
4. Brubaker, 'The "Diaspora" Diaspora'.
5. George Hicks, 1997. *Japan's Hidden Apartheid: The Korean Minority and the Japanese*. London: Ashgate.
6. Digital National Security Archive (DNSA), 1963. 'Department of State, Memorandum of Conversation,' 15 May 1963, http://nsarchive.chadwyck.com.ezproxy.soas.ac.uk/nsa/documents/JU/00233/all.pdf (accessed 24 March 2014).
7. Eberstadt, 'Financial Transfers from Japan to North Korea,' p. 524.
8. Shipper, 'Nationalism of and against Zainichi Koreans,' p. 64.
9. Ryang, *North Koreans in Japan*, p. 124.
10. E. Ben Heine, 2008. 'Japan's North Korean Minority in Contemporary Context'. International Studies BA Thesis, University of Chicago, http://www.atlantic-community.org/index.php/Open_Think_Tank_Article/Japan%27s_North_Koreans (accessed 24 May 2015).
11. Ryang, *North Koreans in Japan*, p. 102.
12. Eberstadt, 'Financial Transfers from Japan to North Korea,' p. 523.
13. Hughes, *Japan's Remilitarisation*, p. 292.
14. Ibid., also see Yasuyo Sakata, 2011. 'Korea and the Japan-US Alliance: A Japanese Perspective,' in Takahashi Inoguchi, G. John Ikenberry and Yochiro Sato (eds), *The US-Japan Security Alliance, Regional Multilateralism*. New York: Palgrave Macmillan; and Yoichi Funabashi, 1999. *Alliance Adrift*. Washington, DC: Council on Foreign Relations.
15. Tessa Morris-Suzuki, 2007. *Exodus to North Korea: Shadows from Japan's Cold War*. Lanham, MD: Rowman & Littlefield.
16. Chanlett-Avery Emma, 2003. *North Korean Supporters in Japan: Issues for US Policy*. United States Congressional Research Service: CRS report number: RL32137, 7 November 2003, https://file.wikileaks.org/file/crs/RL32137.pdf (accessed 2 December 2014).
17. Eberstadt, 'Financial Transfers from Japan to North Korea,' p. 524.
18. Shipper, 'Nationalism of and against Zainichi Koreans,' p. 63.
19. G. H. Han, 2005. *Waga Chosen Soren no tsumi to batsu* (*The Crimes and Punishment of our Chongryun*). Tokyo: Bunshun.
20. Shipper, 'Nationalism of and against Zainichi Koreans,' p. 63; see also Sonia Ryang, 2000. 'The North Korean Homeland of Koreans in Japan,' in Sonia Ryang (ed.), *Koreans in Japan: Critical Voices from the Margin*. London: Routledge.
21. Tsutomu Nishioka, 2003. 'North Korea's Threat to the Japan-US Alliance,' *Japan Echo* 30(2).
22. Agence France Presse, 1999. 'Credit Unions in Japan Suspected of Illegal Remittances to N. Korea,' 29 August 1999.
23. Emma, *North Korean Supporters in Japan*, p. 7.
24. Ibid., p. 5.
25. The Yomiuri Shimbun, 2003. 'Risky Business Leading N. Korea to Ruin,' 22 August 2003, https://database.yomiuri.co.jp/rekishikan/ (accessed 24 May 2015).
26. BBC Monitoring International Reports, 2003. 'Japanese Firm Admits to Exporting Nuke-Related Devices to North Korea,' 9 July 2003.
27. The Yomiuri Shimbun, 2003. 'Government to Monitor N. Korean Ship,' 24 May 2003, https://database.yomiuri.co.jp/rekishikan/ (accessed 24 May 2015).

28. The Yomiuri Shimbun, 2003. 'METI Busts N. Korea Trader,' 19 May 2003, https://database.yomiuri.co.jp/rekishikan/ (accessed 24 May 2015).
29. Ibid.
30. Mindy Kotler, 2003. 'Interdiction may not just modify North Korea's behavior,' NAPSNet Policy Forum, 13 June 2003, http://nautilus.org/napsnet/napsnet-policy-forum/nautilus-institute-policy-forum-online-interdiction-may-not-just-modify-north-koreas-behavior/ (accessed 25 May 2015).
31. Nishioka, 'North Korea's Threat to the Japan-US Alliance'.
32. Foreign Relations of the United States, 1969. 'Memorandum of Conversation,' FRUS 1969–1976. Volume XIX, Part 1, Korea, 1969–1972, National Archives, Nixon Presidential Materials, NSC Files, Box 1023, San Francisco, CA, 21 August 1969, http://history.state.gov/historicaldocuments/frus1969-76v19p1/d35 (accessed 24 May 2015).
33. Hannah Arendt, 1962. *The Origins of Totalitarianism*. Cleveland and New York: World Publishing Co., p. 478.
34. Ibid., p. 11.
35. Louis Fiset and Gail M. Nomura, 2005. 'Introduction,' in Lois Fiset and Gail M. Nomura (eds), *Nikkei in the Pacific Northwest*. Seattle; London: University of Washington Press, p. 11.
36. Ibid., p. 9.
37. Ibid.
38. Mylonas, *The Politics of Nation-Building*, p. 195.

Bibliography

Archival Sources

Central Intelligence Agency
The Digital National Security Archive (DNSA)
The Foreign Relations of the United States, 1945, VI
The Foreign Relations of the United States, 1947, VI
The Foreign Relations of the United States, 1950, V
The Foreign Relations of the United States, 1950, VI
The Foreign Relations of the United States, 1951, VI
The Foreign Relations of the United States, 1958–60, XVIII
National Archives, Nixon Presidential Materials
National Archives and Records Administration (NARA)
National Diet Library

Newspapers and Periodicals

Asahi Newspaper
BBC World News
CCTV News
Chosen Shotokufu kanpo kogai [Chosen Shotokufu newsletter, extra edition]
Congressional Research Service Reports (CRS)
Daily Yomiuri
Dong-A Ilbo
Japan Digest
Japan Times
Korea Herald
The Ministry of Justice: http://www.moj.go.jp
New York Times
Nikkei Weekly
Rodong Shinmun
Sekai Shunho
The White House Office of the Press Secretary
Yomiuri Newspaper

Archival Data

Central Intelligence Agency, 1984. 'Briefing Materials for the President's Meeting with Prime Minister Nakasone, 2 January 1985.' 21 December 1984, Washington, DC:

BIBLIOGRAPHY

http://www.foia.cia.gov/sites/default/files/document_conversions/89801/DOC_0000309669.pdf (accessed 24 March 2014).

Digital National Security Archive (DNSA), 1963. 'Department of State, Memorandum of Conversation,' 15 May 1963, http://nsarchive.chadwyck.com.ezproxy.soas.ac.uk/nsa/documents/JU/00233/all.pdf (accessed 24 March 2014).

Digital National Security Archive (DNSA) Online, 1983. *Defence Estimative Brief, Prospects for South Korea*. 28 March 1983, NLS F95-033/1 #180, DEB-29-83, http://nsarchive.chadwyck.com/quick/displayMultiItemImages.do?Multi=yes&ResultsID=14CDC24B77B&queryType=quick&QueryName=cat&ItemID=CKO00438&ItemNumber=924.

Foreign Relations of the United States, 1947. *The Secretary of War (Patterson) to the Acting Secretary of State*. FRUS 1947, The Far East, Vol. VI, p. 626, 740.00119 Control (Korea)/4-447.

—— 1951. *Undated Memorandum by the Prime Minister of Japan (Yoshida)*. FRUS, 1951, Vol. VI, pp. 833-4.

—— 1951. *The Consultant to the Secretary (Dulles) to the Secretary of State*. 10 February 1951, FRUS, 1951, Vol. VI, pp. 874-80.

—— 1951. *Memorandum Containing the Sections Dealing with Korea from NSO 48/5*. 17 May 1951, FRUS 1951, Vol. VII, Part 1, pp. 439-42, 795-00/5-1751.

—— 1960. *Telegram from the Embassy in Japan to the Department of State*. Foreign Relations of the United States 1958-60, Japan; Korea, Vol. XVIII, p. 422, 16 December 1960.

—— 1969. 'Memorandum of Conversation,' FRUS 1969-1976, Vol. XIX, Part 1, Korea, 1969-1972, National Archives, Nixon Presidential Materials, NSC Files, Box 1023, San Francisco, CA, 21 August 1969, http://history.state.gov/historicaldocuments/frus1969-76v19p1/d35 (accessed 24 May 2015).

Foreign Relations of the United States, Asia and the Pacific (in two parts), 1951. *Memorandum of Conversation, by Mr. Robert A. Fearey of the Office of Northeast Asian Affairs*. FRUS 1951, Vol. VI, pp. 1007-08. 23 April 1951.

Joint Chiefs of Staff, 1945. *Basic Initial Post-Surrender Directive to Supreme Commander for the Allied Powers for the Occupation and Control of Japan (JCS1380/15)*, 3 November 1945, in Government Section, SCAP. *Political Reorientation of Japan* (Washington, DC: US Government Printing Office, 1949), pp. 429-41, http://www.ndl.go.jp/constitution/e/shiryo/01/036/036tx.html#t001 (accessed 15 March 2013).

National Archives and Records Administrations (NARA), 1959. *Telegram from US Ambassador, MacArthur to Secretary of State, Washington*. 7 February 1959, decimal file no. 694.95B/759, parts 1 and 2.

—— 1960. *The Repatriation Problem Revisited* (Confidential, Memorandum), 13 May 1960, Japan and the US, 1960-1976, JU00040, NARA declassified NND959269, http://nsarchive.chadwyck.com/nsa/documents/JU/00040/all.pdf (accessed 15 March 2013).

National Archives, Nixon Presidential Materials, NSC Files, Box 364, Subject Files, National Security Decision Memoranda, Nos 97-144, 1970. *Telegram From the Embassy in Korea to the Department of State*. FRUS 1969-1976, Vol. XIX, Part 1, Korea, 1969-1972, 15 June 1970, http://history.state.gov/historicaldocuments/frus1969-76v19p1/d61 (accessed 2 December 2014).

—— Box 757, Presidential Correspondence 1969-1974, Korea, President Chung Hee Park, 1971. *Letter From President Nixon to Korean President Park*. FRUS 1969-1976, Vol. XIX, Part 1, Korea, 1969-1972, 16 July 1971, http://history.state.gov/historicaldocuments/frus1969-76v19p1/d100 (accessed 2 December 2014).

BIBLIOGRAPHY

Newspapers

Agence France Presse, 1999. 'Credit Unions in Japan Suspected of Illegal Remittances To N. Korea,' 29 August 1999.

Asahi Newspaper, 7 March 1965, http://database.asahi.com/library2e/smendb/d-imageframeset-main.php (accessed 3 August 2015).

—— 12 October 1973, p. 3, http://database.asahi.com/library2e/smendb/d-image-frameset-main.php (accessed 2 December 2014).

—— 6 September 1974, p. 1, http://database.asahi.com/library2e/smendb/d-image-frameset-main.php (accessed 2 December 2014).

—— 17 September 1975, p. 5, http://database.asahi.com/library2e/smendb/d-image-frameset-main.php (accessed 2 December 2014).

—— 22 November 1975, p. 11, http://database.asahi.com/library2e/smendb/d-imageframeset-main.php (accessed 2 December 2014).

—— 10 November 1977, p. 2, http://database.asahi.com/library2e/smendb/d-image-frameset-main.php (accessed 2 December 2014).

—— 4 July 1977, p. 1, http://database.asahi.com/library2e/smendb/d-image-frameset-main.php (accessed 2 December 2014).

—— 9 September 1984, p. 1, http://database.asahi.com/library2e/smendb/d-image-frameset-main.php (accessed 24 March 2014).

—— 31 October 1984, p. 1, http://database.asahi.com/library2e/smendb/d-image-frameset-main.php (accessed 14 March 2014).

—— 20 March 2003, p. 4, http://database.asahi.com/library2e/main/start.php (accessed 15 March 2014).

—— 18 December 2013, p. 1, http://database.asahi.com/library2e/main/start.php (accessed 15 March 2014).

—— 17 December 2013, p. 2, http://database.asahi.com/library2e/main/start.php (accessed 15 March 2014).

BBC Monitoring International Reports, 2003. 'Japanese Firm Admits to Exporting Nuke-Related Devices to North Korea,' 9 July 2003.

BBC World News, 2014. 'N Korea Fires Short-Range Missiles,' 9 July 2014, http://www.bbc.co.uk/news/world-asia-28223183 (accessed 9 July 2014).

Japan Times, 2014. 'Tokyo Hanoi Unite on Defense Ties,' 2 June 2014.

Yomiuri Newspaper, 1955. 'Not overthrowing Japanese Government: North Korean leader's announcement to Zainichi Koreans,' 23 October 1955, https://database.yomiuri.co.jp/rekishikan/viewerMtsStart.action?objectId=l0au26ZbyOEVGbsWuYL7N9K8KjfJOAjWdnw3Vrq5IQg%3D (accessed 2 December 2014).

—— 1959. 'Minority in Japan: Zainichi Korean who Migrated for Survival but 80% of the Unemployment Status,' 14 March 1959, https://database.yomiuri.co.jp/rekishikan/viewerMtsStart.action?objectId=sTE7HhiNcvXGMajuzPfdO2k6LDhVoXJ%2FqDRkf7zEyCg%3D (accessed 2 December 2014).

—— 1959. 'Welcoming for Repatriates in North Korea,' 8 November 1959, https://database.yomiuri.co.jp/rekishikan/viewerMtsStart.action?objectId=ZZ5F9Z8EFJ2hq03JEEZJgWk6LDhVoXJ%2FqDRkf7zEyCg%3D (accessed 2 December 2014).

—— 1960. 'North Korea Repatriation,' 24 November 1960, https://database.yomiuri.co.jp/rekishikan/viewerMtsStart.action?objectId=LHu1psKETCf1SvDlxxxXT3pB6RpTsc7BIXVOYzDte0E%3D (accessed 31 August 2013).

—— 10 November 1965, https://database.yomiuri.co.jp/rekishikan/viewerMtsStart.action?objectId=67R1If%2B%2BbFFI4yQH5DJkvKdmFjPuBLg98YSU4ORfSKM%3D (accessed 3 August 2015).

BIBLIOGRAPHY

—— 1972. 'Tax-Exempt Status Applicable, for Chosen Soren,' 2 July 1972, https://database.yomiuri.co.jp/rekishikan/viewerMtsStart.action?objectId=hESgNt0Pm%2FwJiwph1UJ2VHhu%2BboWjronX%2FW6d9Lc9vU%3D (accessed 2 December 2014).

—— 1974. 'Prison Note of Mun Se Kwang,' 14 December 1974, https://database.yomiuri.co.jp/rekishikan/viewerMtsStart.action?objectId=O9Ljd%2FAOrNrhWaTpJ6Dm9ixuYycY5UrOe9PhRI1SEwc%3D (accessed 2 December 2014).

—— 1975, 'We Strongly Support Revolution in South,' 19 April 1975, http://database.asahi.com/library2e/smendb/d-image-frameset-main.php (accessed 2 December 2014).

—— 22 August 1981, p. 1, https://database.yomiuri.co.jp/rekishikan/viewerMtsStart.action?objectId=NaxrGtBsalz0s01PLdgkCnpB6RpTsc7BIXVOYzDte0E%3D (accessed 2 December 2014).

Yomiuri Shimbun, 2003. 'Risky Business Leading N. Korea to Ruin,' 22 August 2003, https://database.yomiuri.co.jp/rekishikan/ (accessed 24 May 2015).

—— 2003. 'Government to Monitor N. Korean Ship,' 24 May 2003, https://database.yomiuri.co.jp/rekishikan/ (accessed 24 May 2015).

—— 2003. 'METI Busts N. Korea Trader,' 19 May 2003, https://database.yomiuri.co.jp/rekishikan/ (accessed 24 May 2015).

Books and Articles

Adamson, Fiona B., 2002. 'Mobilizing for the Transformation of Home: Politicized Identities and Transnational Practices,' in Nadje Al-Ali and Khalid Koser (eds), *New Approaches to Migration? Transnational Communities and the Transformation of Home*. London: Routledge.

—— 2012. 'Constructing the Diaspora: Diaspora Identity Politics and Transnational Social Movements,' in Terrence Lyons and Peter Mandaville (eds), *Politics from Afar: Transnational Diasporas and Networks*. London: Hurst & Co.

—— 2013. 'Mechanism of Diaspora Mobilization and the Transnationalization of Civil War,' in Jeffrey T. Checkel (ed.), *Transnational Dynamics of Civil War*. Cambridge: Cambridge University Press.

Ahn, Byung-joon, 1983. 'The United States and Korean-Japanese Relations,' in Gerard L. Curtis and Han Sŭng-jŭ (eds), *The U.S.-South Korean Alliance: Evolving Patterns in Security Relations*. Lexington, MA: Lexington Books.

Alex, Delmar-Morgan and Peter Oborne, 2014. 'Why is the Muslim Charity Interpal being Blacklisted as a Terrorist Organization?,' *The Telegraph*, 26 November 2014, http://www.telegraph.co.uk/news/religion/11255294/Why-is-the-Muslim-charity-Interpal-being-blacklisted-as-a-terrorist-organisation.html (accessed 24 May 2015).

Anderson, Benedict, 1991. *Imagined Communities: Reflections on the Origin and Spread of Nationalism*. London, New York: Verso.

—— 1994. 'Exodus,' *Critical Inquiry* 20(2): pp. 314–27.

Appadurai, Arjun, 1996. *Modernity at Large: Cultural Dimensions of Globalization*. Minneapolis: University of Minnesota Press.

Arendt, Hannah, 1962. *The Origins of Totalitarianism*. Cleveland, New York: World Publishing Co.

Armstrong, Charles K., 2014. *The Koreas*. London: Routledge.

Aruga, Tadashi, 1989. 'Security Treaty Revision,' in Iriye Akira, and Warren I. Cohen (eds), *The United States and Japan in the Postwar World*. Lexington, KY: University Press of Kentucky.

Auslin, Michel and Christopher Griffin, 2008. *Securing Freedom: The US-Japanese Alliance in a New Era, A Report of the American Enterprise Institute*. Washington, DC: American Enterprise Institute.

BIBLIOGRAPHY

Barany, George, 1974. 'Magyar Jew or Jewish Magyar?: Reflections on the question of assimilation,' in *Jews and Non-Jews in Eastern Europe, 1918-1945*. New York: Wiley and Sons.
Barth, F., 1998. *Ethnic Groups and Boundaries: The Social Organization of Culture Difference*. Prospect Heights, IL: Waveland Press.
Beasley, William Gerald, 1987. *Japanese Imperialism 1894-1945*. Oxford: Clarendon Press.
Bernstein, Thomas P. and Andrew J. Nathan, 1983. 'The Soviet Union, China and Korea,' in Gerald L. Curtis and Han Sŭng-jŭ, *The U.S.-South Korean Alliance: Evolving Patterns in Security Relations*. Lexington, MA: Lexington Books.
Bhabha, H.K., 1990. 'The Third Space: Interview with Homi Bhabha' in J. Rutherford (ed.), *Identity, Community, Culture, Difference*. London: Lawrence & Wishart.
Bleiker, Roland, 2005. *Divided Korea: Toward a Culture of Reconciliation*. London: University of Minnesota Press.
Brass, Paul R., 1991. *Ethnicity and Nationalism: Theory and Comparison*. London: Sage Publications.
Brazinsky, Gregg, 2007. *Nation Building in South Korea: Koreans, Americans, and the Making of a Democracy*. Chapel Hill: University of North Carolina Press.
Brown, Seyom, 1968. *The Faces of Power*. New York: Columbia University Press.
Brubaker, Rogers, 1996. *Nationalism Reframed*. Cambridge: Cambridge University Press.
—— 2001. 'The Return of Assimilation? Changing Perspectives on Immigration and its Sequels in France, Germany, and the United States,' *Ethnic and Racial Studies* 24(4): pp. 531-48.
—— 2005. 'The "Diaspora" Diaspora,' *Ethnic and Racial Studies* 28(1): 1-19.
—— 2010. 'Migration, Membership, and the Modern Nation-State: Internal and External Dimensions of the Politics of Belonging,' *Journal of Interdisciplinary History* XLI, I: pp. 61-78.
Bryant, Nick, 2015. 'The Decline of US Power?,' BBC News, 10 July 2015, http://www.bbc.co.uk/news/world-us-canada-33440287 (accessed 19 July 2015).
Burr, William, 1998. *The Kissinger Transcripts*. New York: The New Press.
Byman, Daniel, Peter Chalk, Bruce Hoffman, William Rosenau, and David Brannan (eds), 2001. *Trends in Outside Support for Insurgent Movements*. National Security Research Division: RAND Corporation.
Calder, Kent E., 1988. 'Japanese Foreign Economic Policy Formation: Explaining the Reactive State,' *World Politics* 40(4): pp. 517-41.
—— 2007. 'Beneath the Eagle's Wings? The Political Economy of Northeast Asian Burden-Sharing in Comparative Perspective,' *Asian Security* 2(3): pp. 148-73.
Caprio, Mark, 2008. 'The Cold War Explodes in Kobe – The 1948 Korean Ethnic School Riots and US Occupation Authorities,' *The Asia-Pacific Journal: Japan Focus* Vol. 48-2-08.
—— 2009. *Japanese Assimilation Policies in Colonial Korea, 1910-1945*. Seattle: University of Washington Press.
CCTV News, 2014. 'Diplomatic Talks between S.Korea, Japan End Fruitless' 3 December 2014, http://english.cntv.cn/20140312/105470.shtml (accessed 2 December 2014).
Cha, Victor D., 1999. *Alignment despite Antagonism: The United States-Korea-Japan Security Triangle*. Stanford: Stanford University Press.
—— 2003. 'Japan-Korea Relations: Contemplating Sanctions,' *Comparative Connections* 5(1).
—— 2011. '"Rhee-strait": The Origins of the U.S.-ROK Alliance,' *International Journal Korean Studies* 15(1): pp. 1-15.
Chapman, David, 2008. *Zainichi Korean Identity and Ethnicity*. London, New York: Routledge.
Chang, Yunshik, 1971. 'Colonization as Planned Change: The Korean Case,' *Modern Asian Studies* 5(2): pp. 161-86.

BIBLIOGRAPHY

Charles, King and Neil J. Melvin, 1990/2000. 'Diaspora Politics: Ethnic Linkages, Foreign Policy and Security in Eurasia,' *International Security* 24(3): pp. 108–38.

Checkel, Jeffrey T., 2013. 'Transnational Dynamics of Civil War,' Jeffrey T. Checkel (ed.), in *Transnational Dynamics of Civil War*. Cambridge: Cambridge University Press.

Cheong, Sung-Hwa, 1991. *The Politics of Anti-Japanese Sentiment in Korea: Japanese-South Korean Relations Under American Occupation, 1945–1952*. New York: Greenwood Press.

Ching, Leo T.S., 2001. *Becoming 'Japanese': Colonial Taiwan and the Politics of Identity Formation*. London: University of California Press.

Ch'oe, In-bŏm, 2003. 'Korean Diaspora in the Making: Its Current Status and Impact on the Korean Economy,' in C. Fred Bergsten and In-bŏm Ch'oe (eds), *The Korean Diaspora in the World Economy*. Washington, DC: Institute for International Economics, 2003.

Cho, Soon Sung, 1967. *Korea in World Politics, 1940–1950*. Berkeley: University of California Press.

Chŏng, Sŏng-hwa, 1991. *The Politics of Aanti-Japanese Sentiment in Korea: Japanese-South Korean Relations Under American Occupation, 1945–1952*. New York: Greenwood Press.

Chōsen Sōren, 2005. *Chōsen Sōren*. Tokyo: Chōsen Sōren.

Chosen Sotokufu, 1932. *Tokei nempo* [Korean Government-General], Seoul.

—— 1938. *Chosen Shotokufu kanpo kogai* [Chosen Shotokufu newsletter, extra edition] 1938:2.

Chou, Wan-yao, 1996. 'The Kominka Movement in Taiwan and Korea: Comparisons and Interpretation,' in Ramon Hawley Myers, Mark R. Peattie and Ching-chih Chen (eds), *The Japanese Wartime Empire, 1931–1945*. Princeton, NJ: Princeton University Press.

Chung, Chae-ho, 2007. *Between Ally and Partner: Korea-China Relations and the United States*. New York: Columbia University Press.

Chung, Erin Aeran, 2010. *Immigration and Citizenship in Japan*. New York: Cambridge University Press.

Chung, Kyungmo, 1975. 'Aru Minzokushugisha No Shogai,' (A life of Nationalist) in *Sekai*, December 1975.

Clark, Donald N., 1980. *The Kwangju Regime in South Korea*. Boulder: Westview Press.

Cliff, Roger, Mark Burles, Michael S. Chase, Derek Eaton and Kevin L. Pollpeter, 2007. *Entering the Dragon's Lair: Chinese Antiaccess Strategies and Their Implications for the United States*. Santa Monica, CA: Rand Corporation.

Clifford, James, 1994. 'Diasporas,' *Cultural Anthropology* 9(3): pp. 302–38.

Clifford, Mark L., 1998. *Troubled Tiger*. New York: An East Gate Book.

Clogg, Richard (ed.), 2003. *Minorities in Greece: Aspects of a Plural Society*. London: C. Hurst.

Cochrane, F., 2007. 'Civil Society Beyond the State: The Impact of Diaspora Communities on Peace Building,' *Global Media Journal*, Mediterranean Edition 2(2): pp. 19–29.

Colbert, Evelyn, 1986. 'Japan and Republic of Korea,' *Asian Survey* 26(3): pp. 273–91.

Colley, Linda, 1996. *Britons: Forging the Nation 1707–1837*. London: Vintage.

Collier, Paul, 2000. *Economic Causes of Civil Conflict and their Implications for Policy*. Washington, DC: World Bank.

Cossa, Ralph A., 1999. 'Preface,' in Ralph A. Cossa (ed.), *US-Korea-Japan Relations: Building Toward a Virtual Alliance*. Washington, DC: The CSIS Press.

Cox, Robert, 1989. 'Middlepowermanship, Japan and Future World Order,' *International Journal* 44(4): pp. 823–62.

Cumings, Bruce, 1981. *The Origins of the Korean War: Liberation and the Emergence of Separate Regimes, 1945–1947*. Princeton, NJ: Princeton University Press.

BIBLIOGRAPHY

—— 1984. 'The Legacy of Japanese Colonialism in Korea,' in Ramon Hawley Myers, Mark R. Peattie and Ching-chih Chen (eds), *The Japanese Colonial Empire, 1895-1945*. Princeton, NJ: Princeton University Press.
—— 1997. *Korea's Place in the Sun: A Modern History*. London: W.W. Norton & Company.
Curtis, Gerald L. and Sŭng-jŭ Han, 1983. *The U.S.-South Korean Alliance: Evolving Patterns in Security Relations*. Lexington, MA: Lexington Books.
Darwin, John, 1988. *Britain and Decolonization*. New York: St. Martin's.
Deutsch, Karl W., 1968. 'The Trend of European Nationalism – The Language Aspect,' in Joshua A. Fishman (ed.), *Readings in the Sociology of Language*. The Hague: Mouton.
De Vos, George A. and William O. Wetherall, 1974. *Japan's Minorities: Burakumin, Koreans, Ainu*, new edn. London: Minority Rights Group.
Dower, J.W., 1979. *Empire and Aftermath: Yoshida Shigeru and the Japanese Experience, 1878-1954*. Cambridge, MA: Harvard University Press.
Duus, Peter, 1996. 'Introduction,' in Peter Duus, Ramon Hawley Myers, Mark R. Peattie and Wan-yao Chou (eds), *The Japanese Wartime Empire, 1931-1945*. Princeton, NJ: Princeton University Press.
DW AKADEMIE, 2014. 'Chinese President Xi visits Seoul in apparent snub to North Korea,' 3 July 2014, http://www.dw.de/chinese-president-xi-visits-seoul-in-apparent-snub-to-north-korea/a-17754872 (accessed 9 July 2014).
Easley, Leif-Eric, 2008. 'Securing Japan's Positive Role in North-South Reconciliation: The Need for a Strong US-ROK Alliance to Reassure Japan,' *Academic Paper Series on Korea* (1): pp. 171-4.
Eberstadt, Nicholas, 1996. 'Financial Transfers from Japan to North Korea: Estimating the Unreported Flows,' *Asian Survey* 36(5): pp. 523-42.
Eberstadt, Nicholas, Aaron L. Friedberg and Geun Lee (eds), 2008. 'Introduction: What If? A World Without the US-ROK Alliance,' *Asia Policy* (5): pp. 2-5.
Eckert, Carter J., 1996. 'Total War, Industrialization and Social Change in Late Colonial Korea,' in Ramon Hawley Myers, Mark R. Peattie and Ching-chih Chen (eds), *The Japanese Wartime Empire, 1931-1945*. Princeton, NJ: Princeton University Press.
Emma, Chanlett-Avery, 2003. *North Korean Supporters in Japan: Issues for US Policy*. United States Congressional Research Service. CRS report number: RL32137, 7 November 2003, https://file.wikileaks.org/file/crs/RL32137.pdf (accessed 2 December 2014).
Emmerson, John K., 1978. *The Japanese Thread: A Life in the US Foreign Service*. New York: Holt, Rinehart and Winston.
Farrar-Hockley, Anthony, 1990. *The British Part in the Korean War*. London: HMSO.
Finnegan, Michael, 2009. *Managing Unmet Expectations in the US-Japan Alliance*. The National Bureau Asian Research: NBR Special Report #17.
Fiset, Louis and Gail M. Nomura, 2005. 'Introduction,' in Louis Fiset and Gail M. Nomura (eds), *Nikkei in the Pacific Northwest*. Seattle and London: University of Washington Press.
Fuji, Kamiya, 1982. 'The Northern Territories: 130 Years of Japanese Talks with Czarist Russia and the Soviet Union,' in Donald S. Zagoria (ed.), *Soviet Policy in East Asia*. New Haven and London: Yale University Press.
Fujii, Hyogo, 1965. 'Nikkan Keizai Kyoroku heno Kozo' (Ideas on Japan-Korean Economic Cooperation), *Jiyu*.
Fujishima, Udai, 1966. 'Waga Michio Yuku Jishin' (Confidence of Going My Own Way), Economisto, Vol. XLIV.
Fukuoka, Yasunori, 1993. *Zainichi Kankoku-Chosenjin*. Tokyo: Chuo Koronsha.
Funabashi, Yoichi, 1999. *Alliance Adrift*. Washington, DC: Council on Foreign Relations.

BIBLIOGRAPHY

―――― 2007. *The Peninsula Question: A Chronicle of the Second Nuclear Crisis*. Washington, DC: Brookings Institution.
Gaddis, John Lewis, 1972. *The US and the Origins of the Cold War*. New York: Columbia University Press.
―――― 1977. 'Korea in American Politics, Strategy, and Diplomacy, 1945–50,' in Yōnosuke Nagai and Iriye Akira (eds), *The Origins of the Cold War in Asia*. Tokyo: University of Tokyo Press.
Gallicchio, Marc, 1991. 'The Kuriles Controversy: U.S. Diplomacy in the Soviet-Japan Border Dispute, 1941–1956,' *Pacific Historical Review* 60(1): pp. 69–101.
Gardner, Paul F., 1999. *The Nakasone Government: Prospects and Problems*. United States Department of State Review, Bureau of Intelligence and Research: Issues Paper, Date/Case ID: 8 September 1999, 199501403, http://nsarchive.chadwyck.com/nsa/documents/JA/01059/all.pdf (accessed 22 March 2014).
Gellner, Ernest, 2006. *Nations and Nationalism*, 2nd edn. Malden, MA: Blackwell Publisher.
George, Alexander L. and Andrew Bennett, 2005. *Case Studies and Theory Development in the Social Sciences*. Cambridge: The MIT Press.
Giddens, Anthony, 1979. *Central Problems in Social Theory: Action, Structure and Contradiction in Social Analysis*. London: Macmillan.
Gilroy, Paul, 1993. *The Black Atlantic: Modernity and Double Consciousness*. Cambridge, MA: Harvard University Press.
Global Reporting Network Publications, 1998. 'Politics and the Agreed Framework: North Korean Deal On Thin Ice?,' *Global Beat Issue Brief*, No. 44, 10 November 1998, http://www.bu.edu/globalbeat/pubs/ib44.html (accessed 24 May 2015).
Gluck, Carol, 2007. 'Operations of Memory,' in Sheila Miyoshi Jager and Rana Mitter (eds), *Ruptured Histories: War, Memory and the Post-Cold War in Asia*. London: Harvard University Press.
Green, Michael Jonathan, 2001. *Japan's Reluctant Realism*. New York: Palgrave.
Griffin, Christopher and Michael Auslin, 2008. 'Time for Trilateralism?,' *Asian Outlook* (2). Washington, DC: American Enterprise Institute for Public Policy Research.
Hah, Chong-do and Martin Jeffrey, 1975. 'Toward a Synthesis of Conflict and Integration Theories of Nationalism,' *World Politics* 27(3): pp. 361–86.
Hahei Chekku, Henshū Iinkai, 1997. *Kore Ga Beigun e No 'Omoiyari Yosan' Da!: 'Nichi-Bei Anpo'-dokuhon*. Shohan. Tōkyō: Shakai Hyōronsha.
Hall, Stuart, 1990. 'Cultural Identity and Diaspora,' in J. Rutherford (ed.), *Identity: Community, Culture, Difference*. London: Lawrence & Wishart.
Hammar, Tomas, 1990. *Democracy and the Nation State: Aliens, Denizens, and Citizens in a World of International Migration*. Aldershot: Avebury.
Han, Enze, 2013. *Contestation and Adaptation*. Oxford: Oxford University Press.
Han, G.H., 2005. *Waga Chosen Soren no tsumi to batsu* (The crimes and punishment of our Chongryun). Tokyo: Bunshun.
Han, Sung-joo, 1983. 'South Korea and the United States: Past, Present, and Future,' in Curtis Gerald L. and Han Sung-joo (eds), *The U.S.-South Korean Alliance: Evolving Patterns in Security Relations*. Toronto: Heath and Company.
Havens, Thomas R.H., 1987. *Fire across the Sea: The Vietnam War and Japan, 1965–1975*. Princeton, NJ: Princeton University Press.
Hechter, Michael, 1975. *Imagined Colonialism: The Celtic Fringe in British National Development, 1536–1966*. Berkeley: University of California Press.
Heine, E. Ben, 2008. 'Japan's North Korean Minority in Contemporary Context,' International Studies BA Thesis, The University of Chicago, http://www.atlantic-community.org/index.php/Open_Think_Tank_Article/Japan%27s_North_Koreans (accessed 24 May 2015).

BIBLIOGRAPHY

Helder, De Schutter and Ypi Lea, 2015. 'The British Academy Brian Barry Prize Essay: Mandatory Citizenship for Immigrants,' *British Journal of Political Science* 45(2): pp. 235–51.

Hicks, George, 1997. *Japan's Hidden Apartheid: The Korean Minority and the Japanese.* London: Ashgate.

Hobsbawm, E.J., 1997. *The Age of Empire, 1875–1914.* London: Abacus.

Hongo, Jun, 2007. 'Court Rules Chongryun Property not Tax-Exempt,' *Japan Times*, 21 July 2007, http://www.japantimes.co.jp/news/2007/07/21/national/court-rules-chongryun-property-not-tax-exempt/#.VWcujbFwaM8 (accessed 15 March 2014).

Horowitz, Donald L., 1991. 'Irredentas and Secessions: Adjacent Phenomena, Neglected Connections,' in Naomi Chazan (ed.), *Irredentism and International Politics.* Boulder, CO: London.

—— 1993. 'Democracy in Divided Societies,' *Journal of Democracy* 4(4): pp. 18–38.

Hugh, Seton-Watson, 1977. *Nations and States: An Enquiry into the Origins of Nations and the Politics of Nationalism.* London: Methuen.

Hughes, Christopher W., 2009. *Japan's Remilitarisation.* London, New York: International Institute for Strategic Studies, Routledge.

—— 2009. 'Super-Sizing the DPRK Threat: Japan's Evolving Military Posture and North Korea,' *Asian Survey* 49(2): pp. 291–311.

Hwang, Kyung Moon, 2010. *A History of Korea: An Episodic Narrative.* Hampshire: Palgrave Macmillan.

Ina, Hisayoshi, 1996. 'Fear of US-China Ties Rests on Flawed Premises,' *Nikkei Weekly*, 18 December 1996.

Inokuchi, Hiromitsu, 2000. 'Korean Ethnic Schools in Occupied Japan, 1945–52,' in Sonia Ryang (ed.), *Koreans in Japan: Critical Voices from the Margin.* New York: Routledge.

Istin, Engin F. and Patricia K. Wood, 1999. 'Redistribution, Recognition, Representation,' Engin F. Istin and Patricia K. Wood (eds), in *Citizenship and Identity.* London: Sage Publications.

I-te-Chen, Edward, 1984. 'The Attempt to Integrate the Empire, 1895–1945,' in Ramon Hawley Myers, Mark R. Peattie, and Ching-chih Chen (eds), *The Japanese Colonial Empire, 1895–1945.* Princeton, NJ: Princeton University Press.

Japan Red Cross Society, 1956. *The Repatriation Problem of Certain Koreans Residing in Japan.* Tokyo: Japan Red Cross Society.

Japanese Ministry of Foreign Affairs, 1953. *Komura Gaiko Shi* (History of Komura's Diplomacy) *Nihon Gaiko Bunsho*, Tokyo, 1.

—— 1954. *Komura Gaiko Shi* (History of Komura's Diplomacy) *Nihon Gaiko Bunsho*, Tokyo, 29.

Jenne, E.K., 2007. *Ethnic Bargaining: The Paradox of Minority Empowerment.* Ithaca: Cornell University Press.

—— 2008. *Group Demands as Bargaining Positions: Signals, Cues and Minority Mobilization in East Central Europe.* UMI Number: 9995233.

John, Swenson-Wright, 2014. 'What Japan's Military Shift Means,' BBC World News, 2 July 2014, http://www.bbc.co.uk/news/world-asia-28122791 (accessed 9 July 2014).

Johnson, Bruce F., 1953. *Japanese Food Management in World War II.* Stanford: Stanford University Press.

Kahler, Miles, 1997. 'Empires, Neo-Empires and Political Change,' in K. Dawisha and B. Armonk Parrott, *The End of Empire: The Transformation of the USSR in Comparative Perspective.* New York: M.E. Sharpe.

Kamihara, Noboru, 1939. *Komin shinmin ikusei no genjo* [The Present State of Bringing Up Imperial Subjects], in *Chosen*, pp. 39–48.

BIBLIOGRAPHY

Kang, David, 2007. *China Rising*. New York: Columbia University Press.
Kang, Hildi, 2001. *Under the Black Umbrella: Voices from Colonial Korea, 1910–1945*. Ithaca, London: Cornell University Press.
Kang, Sang-Jung, 1991. '*Hoho toshiteno zainichi* [Zainichi as method – in response to Mr Yang Tae- Ho's critique], Osaka, Kaifusha.
—— 1996. *Orientarizumu no kanatani*. Tokyo: Kodansha.
—— 2004. *Zainichi*. Tokyo: Kodansha.
—— 2011. 'Tunnelling Through Nationalism: The Phenomenology of a Certain Nationalist,' *The Asia-Pacific Journal* 9, 36(2).
Kaplan, Stephen S., 1981. *Diplomacy of Power: Soviet Armed Forces as a Political Instrument*. Washington, DC: Brookings Institution.
Kashiwazaki, Chikako, 2000. 'The Politics of Legal Status,' in Sonia Ryang (ed.), *Koreans in Japan: Critical Voices from the Margin*. London: Routledge.
—— 2009. 'The Foreigner Category for Koreans in Japan,' in Sonia Ryang and John Lie (eds), *Diaspora without Homeland*. London: University of California Press.
KEDO Press, 1998. 'KEDO Executive Board Agrees on Cost-Sharing for Light Water Reactor Project,' 10 November 1998.
Keohane, Robert O., 1978. 'American Policy and Trade-Growth Struggle,' *International Security* 3(3): pp. 20–43.
—— 1984. *After Hegemony: Cooperation and Discord in the World Political Economy*. Princeton: Princeton University Press.
Keohane, Robert O. and Joseph S. Nye, 1977. *Power and Interdependence: World Politics in Transition*. Boston: Little Brown.
Kerr, G.H., 1966. *Formosa Betrayed*. London: Eyre & Spottiswoode.
Khrushchev, Nikita, 1974. *Khrushchev remembers; the last testament*, trans. and ed. Strobe Talbott. London: Andre Deutsch
Khrushchev, N.S. and S. Talbott, 1974. *Khrushchev Remembers; The Last Testament*. London: Andre Deutsch.
Kim, Bumsoo, 2006. 'From Exclusion to Inclusion? The Legal Treatment of "Foreigners" in Contemporary Japan,' *Immigrants & Minorities* (1): pp. 51–73.
—— 2011. 'Changes in the Socio-Economic Position of Zainichi Koreans: A Historical Overview,' *Social Science Japan Journal* 14(2): pp. 233–45.
Kim, Hong Nak, 1982. 'After the Park Assassination,' *Journal of Northeast Asian Studies* 1(4): pp. 71–90.
—— 1987. 'Japanese-Korean Relations in the 1980s,' *Asian Survey* 27(5): pp. 497–514.
Kim, Hyung Kook, 1995. *The Division of Korea and the Alliance Making Process: Internationalization of Internal Conflict and Internalization of International Struggle, 1945–1948*. London: University Press of America.
Kim, Joungwon Alexander, 1970. 'Soviet Policy in North Korea,' *World Politics* 22(2): pp. 237–54.
Kim, Marie Seong-Hak, 2012. *Law and Custom in Korea: Comparative Legal History*. Cambridge: Cambridge University Press.
Kim, Seung-Young, 2009. *American Diplomacy and Strategy toward Korea and Northeast Asia, 1882–1950 and After*. New York: Palgrave Macmillan.
—— 2012. 'Balancing Security Interest and the "Mission" to Promote Democracy: American Diplomacy toward South Korea since 1969,' in Robert A. Wampler (ed.), *Trilateralism and Beyond: Great Power Politics and the Korean Security Dilemma During and Aafter the Cold War*. Kent, OH: Kent State University Press.
—— 2013. 'Miki Takeo's Initiative on the Korean Question and U.S-Japanese Diplomacy, 1974–1976,' *Journal of American-East Asian Relations* 20: pp. 377–405.
Kim, Tong Myung, 1988. *Zainichi chosenjin no daisan no michi*. Tokyo: Kaifusha.

King, Charles and Neil J. Melvin, 1990/2000. 'Diaspora Politics: Ethnic Linkages, Foreign Policy and Security in Eurasia,' *International Security* 24(3): pp. 108–38.

Ko, Mika, 2010. *Japanese Cinema and Otherness: Nationalism, Multiculturalism and the Problem of Japaneseness*. London: Routledge.

Kodawau, Yusaku, 1938. *Kokoku Shinmin Taruno Chigaku No Dettei Suru Chosen Kyoikurei No Gaisetsu*, in *Bunkyo No Chosen*, March edition.

Kotler, Mindy, 2003. 'Interdiction may not Just Modify North Korea's Behavior,' NAPSNet Policy Forum, 13 June 2003, http://nautilus.org/napsnet/napsnet-policy-forum/nautilus-institute-policy-forum-online-interdiction-may-not-just-modify-north-koreas-behavior/ (accessed 25 May 2015).

Kou, Y., 2001. 'Nihon Kokuseki todokede hoan to zainichi korian no sentaku,' (The proposal for an automatic access to Japanese nationality and resident Koreans' choices), *Sekai*: pp. 169–76.

Krasner, Stephen D., 1986. 'Trade Conflicts and the Common Defence: The United States and Japan,' *Political Science Quarterly* 101(5): pp. 787–806.

Ku, Dae-yeol, 1985. *Korea under Colonialism: The March First Movement and Anglo-Japanese Relations*. Seoul: Seoul Computer Press.

Lankov, Andrei, 2005, *Crisis in North Korea: The Failure of De-Stalinization, 1956*, Honolulu, University of Hawaii Press.

—— 2010. 'Forgotten People: The Koreans of Sakhalin Island in 1945–1991,' www.nkeconwatch.com/nk-uploads/Lankov-Sakhalin-2010.pdf (accessed 24 May 2015).

Lee, Chae-Jin, 2006. *A Troubled Peace: U.S. Policy and the Two Koreas*. Baltimore: Johns Hopkins University Press.

Lee, Chae-Jin and Doo-Bok Park, 1996. *China and Korea: Dynamic Relations*. Stanford, CA: Hoover Institution Press.

Lee, Chae-Jin and Hideo Sato, 1982. *U.S. Policy toward Japan and Korea: A Changing Influence Relationship*. New York: Praeger.

Lee, Changsoo and George A. De Vos, 1981. *Koreans in Japan: Ethnic Conflict and Accommodation*. Berkeley: University of California Press.

Lee, Chong-Sik, 1985. *Japan and Korea: The Political Dimension*. Stanford, CA: Hoover Institution Press.

Lee, Joyman, 2011. 'Islands of Conflict,' *History Today* 61(5): pp. 24–6.

Lee, Won-sul, 1982. *The United States and the Division of Korea, 1945*. Seoul: Kyunghee University Press.

Leeds, Brett Ashley and Savun Burcu, 2007. 'Terminating Alliances: Why Do States Abrogate Agreements?,' *The Journal of Politics* 69(4): pp. 1118–32.

Li, Luye, 1991. 'The Current Situation in Northeast Asia: A Chinese View,' *Journal of Northeast Asian Studies* 10(1): pp. 78–81.

Lie, John, 2001. *Multiethnic Japan*. London: Harvard University Press.

—— 2008. *Zainichi (Koreans in Japan): Diasporic Nationalism and Postcolonial Identity*. London: University of California Press.

Lim, Youngmi, 2009. 'Korean Roots and Zainichi Routes,' in Sonia Ryang and John Lie (eds), *Diaspora without Homeland*. London: University of California Press.

Lind, Jennifer, 2008. *Sorry States*. Ithaca and London: Cornell University Press.

Liska, G., 1962. *Nations in Alliance: The Limits of Interdependence*. Baltimore, MD: Johns Hopkins University Press.

Loughlin, J.O. and H. Heske, 1991. 'From "Geopolitik to Geopolitique": Converting a Discipline for War to a Discipline for Peace,' in *The Political Geography of Conflict and Peace*. London: Belhaven.

Lowe, P., 1986. *The Origins of the Korean War*. London: Longman Group Limited.

BIBLIOGRAPHY

Manabe, Hiroki, 2013. 'INTERVIEW/ Carol Gluck: Change in Japan is a long-distance run,' *Asahi Newspaper*, 17 September 2013, http://ajw.asahi.com/article/views/opinion/AJ201309170004 (accessed 14 March 2014).

Manela, E., 2007. *The Wilsonian Moment: Self-Determination and the International Origins of Anticolonial Nationalism*. New York and Oxford: Oxford University Press.

Mann, Michael, 2005. *The Dark Side of Democracy: Explaining Ethnic Cleansing*. New York: Cambridge University Press.

Miller, Benjamin, 2007. *States, Nations, and the Great Powers: The Sources of Regional War and Peace*. Cambridge and New York: Cambridge University Press.

—— 2002. *When Opponents Cooperate: Great Power Conflict and Collaboration in World Politics*, second edn. Ann Arbor: University of Michigan Press.

Miller, Robert H., 2001. *Nakasone's Visit to Washington: The View from Japan*, United States Department of State Review Authority:, Date/Case ID: 5 June 2001, 199504020, http://nsarchive.chadwyck.com/nsa/documents/JA/01400/all.pdf (accessed 22 March 2014).

Ministry of Foreign Affairs of Japan, 1953. *Komura Gaiko Shi* (History of Komura's Diplomacy) Tokyo, vol. 1, *Nihon Gaiko Bunsho*.

—— 1954. *Komura Gaiko Shi* (History of Komura's Diplomacy) Tokyo, vol. 29, *Nihon Gaiko Bunsho*.

—— 1998. *Japan-Republic of Korea Joint Declaration: A New Japan-Republic of Korea Partnership toward the Twenty-First Century*. 8 October 1998, http://www.mofa.go.jp/region/asia-paci/korea/joint9810.html (accessed 15 March 2014).

—— 2002. *Speech by Prime Minister of Japan Junichiro Koizumi Japan and ASEAN in East Asia- A Sincere and Open Partnership*. Singapore, 14 January 2002, http://www.mofa.go.jp/region/asia-paci/pmv0201/speech.html (accessed 15 March 2014).

———2002. *Prime Minister Junichiro Koizumi's Visit to North Korea*. 17 September 2002, http://www.mofa.go.jp/region/asia-paci/n_korea/pmv0209/ (accessed 6 June 2015).

Mitchell, Richard H., 1967. *The Korean Minority in Japan*, first edn. California: California University Press.

Mizuno, Naoki, 2004. 'Chosenjin no namae to shokuminchi-shugi,' [Names of Koreans and Colonial Rule], in Naoki Mizuno (ed.), *Seikatsu no naka no shokuminchi-shugi*. Kyoto, *Jimbun Shoin*: pp. 35–77.

Morley, James William, 1983. 'The Dynamics of the Korean Connection,' in Gerald L. Curtis and Sŭng-jŭ Han (eds), *The U.S.-South Korean Alliance: Evolving Patterns in Security Relations*. Lexington, MA: Lexington Books.

Morris-Suzuki, Tessa, 2007. *Exodus to North Korea: Shadows from Japan's Cold War*. Lanham, MD: Rowman & Littlefield Publishers.

—— 2010. *Borderline Japan: Frontier Controls, Foreigners and the Nation in the Postwar Era*. New York: Cambridge University Press.

Morrow, James D., 1986. 'A Spatial Model of International Conflict,' *American Political Science Review*, 80(4): pp. 1131–50.

—— 1991. 'Alliances and Asymmetry: An Alternative to the Capability Aggregation Model of Alliances,' *American Journal of Political Science* 35(4): pp. 904–33.

—— 1994a. 'Alliances, Credibility, and Peacetime Costs,' *Journal of Conflict Resolution* (2): pp. 270–97.

—— 1994b. *Game Theory for Political Scientists*. Princeton, NJ: Princeton University Press.

Myers, Ramon H., 1984. 'Post World War II Japanese Historiography of Japan's Formal Colonial Empire', in Ramon H. Myers and Mark R. Peattie (eds), *The Japanese Colonial Empire: 1895–1945*. Princeton: Princeton University Press.

Mylonas, Harris, 2012. *The Politics of Nation-building: Making Co-Nationals, Refugees, and Minorities*. Cambridge: Cambridge University Press.

BIBLIOGRAPHY

—— 2013. 'The Politics of Diaspora Management in the Republic of Korea,' *The Asian Institute for Policy Studies*, Issue Brief No. 81, 20 November.
Narushige, Michitaka, 2009. *North Korea's Military and Diplomatic Campaigns, 1966–2008*. New York: Routledge.
Newman, David, 2000. 'Citizenship, Identity and Location' in Klaus Dodds and David Atkinson (eds), *Geopolitical Ttraditions: A Century of Geopolitical Thought*. London: Routledge.
Nicholas, Eberstadt, 1996. 'Financial Transfers from Japan to North Korea: Estimating the Unreported Flows,' *Asian Survey* 36(5): pp. 523–42.
Nicholas, Eberstadt, Aaron L. Friedberg and Geun Lee, 2008. 'Introduction: What If? A World without the US-ROK Alliance,' *Asia Policy* (5).
Nishioka, Tsutomu, 2003. 'North Korea's Threat to the Japan-US Alliance,' *Japan Echo* 30(2).
Nixon, Richard, 1967. 'Asia After Vietnam,' *Foreign Affairs*, October 1967, https://www.foreignaffairs.com/articles/asia/1967-10-01/asia-after-viet-nam (accessed 23 May 2015).
Oberdorfer, Don, 2001. *Two Koreas*. New York: Basic Books.
Office of the Secretary of Defense, 2008. *Annual Report to Congress: Military Power of the People's Republic of China*, 5, www.defenselink.mil/pubs/pdfs/China_Military_Report_08.pdf (accessed 9 July 2014).
Oh, Ingyu, 2011. 'New Politico-Cultural Discourse in East Asia?,' in Marie Soderberg (ed.), *Changing Power Relations in Northeast Asia*. London: Routledge.
Okonogi, Masao, 1977. 'The Domestic Roots of the Korean War' in Nagai Yonosuke and Iriye Akira (eds), *The Origins of the Cold War in Asia*. Tokyo: Tokyo University Press.
—— 1985. 'A Japanese Perspective on Korea-Japan Relations' in Chin-Wee Chung, Ky-Moon Ohm, Suk-Ryil Yu, and Dal-Joong Chnag (eds), *Korea and Japan in World Politics*. Seoul: Korean Association of International Relations.
Onuf, Nicholas Greenwood, 1989. *World of our Making: Rules and Rule in Social Theory and International Relations*. Columbia: University of South Carolina Press.
Onuma, Yasuaki, 1979. 'Shutsunyukoku kanri hosei no seiritsukatei,' (The process of the Establishment of the Immigration Control System) 14, *Horitsu jiho* 51(5): pp. 100–106.
Pak, Soon-Yong and Keumjoong Hwang, 2011. 'Assimilation and Segregation of Imperial Subjects: "Educating" the Colonised during the 1910–1945 Japanese Colonial Rule of Korea,' *Paedagogica Historica* 47(3): pp. 377–97.
Park, Il, 1999. *'Zainichi to iu ikikata, (A Way of life as zainichi)'*. Tokyo: Kodansha.
Park, Jae Jeok, 2011. 'The US-led Alliances in the Asia-Pacific: Hedge against Potential Threats or an Undesirable Multilateral Security Order?,' *The Pacific Review* 24(2): pp. 137–58.
Park, Jung-Sun and Paul Y. Chang, 2005. 'Contention in the Construction of a Global Korean Community: The Case of the Overseas Korean Act,' *The Journal of Korean Studies* 10(1): pp. 1–27.
Peattie, Mark R., 1984. 'Introduction,' in Ramon Hawley Myers, Mark R. Peattie and Ching-chih Chen (eds), *The Japanese Colonial Empire, 1895–1945*. Princeton, NJ: Princeton University Press.
—— 1984. 'Japanese Attitudes toward Colonialism,' in Ramon Hawley Myers, Mark R. Peattie and Ching-chih Chen (eds), *The Japanese Colonial Empire, 1895–1945*. Princeton, NJ: Princeton University Press.
Pratt, Cranford, 1990. *Middle Power Internationalism: The North-South Dimension*. Kingston and Buffalo: McGill-Queen's University Press.
Ramon, Myers, 1984. 'Post World War II Japanese Historiography of Japan's Formal Colonial Empire,' in Ramon Hawley Myers, Mark R. Peattie and Ching-chih Chen (eds), *The Japanese Colonial Empire, 1895–1945*. Princeton, NJ: Princeton University Press.
Reischauer, Edwin O., 1967. 'Our Dialogue with Japan,' *Foreign Affairs*, pp. 215–28.

BIBLIOGRAPHY

Robert, J. Watson, 1986. *History of the Joint Chiefs of Staff*, vol. 5: The JCS and National Policy, 1952–1954. Washington, DC: The Library of Congress, pp. 267–75.
Ryang, Sonia, 1997. *North Koreans in Japan*. Boulder, CO: Westview Press.
—— 2000. 'The North Korean homeland of Koreans in Japan,' in Sonia Ryang (ed.), *Koreans in Japan: Critical Voices from the Margin*. London: Routledge.
—— 2008. *Writing Selves in Diaspora*. New York: Lexington Books.
—— 2009. 'Introduction,' in Sonia Ryang and John Lie (eds), *Diaspora without Homeland: Being Korean in Japan*. London: University of California Press.
—— 2009. 'Visible and Vulnerable,' in Sonia Ryang and John Lie (eds), *Diaspora without Homeland: Being Korean in Japan*. London: University of California Press.
Safran, William, 1991. 'Diasporas in Modern Societies: Myths of Homeland and Returns,' *Diaspora: A Journal of Transnational Studies* 1(1): pp. 83–99.
Said, Edward, 1978. *Orientalism*. New York: Pantheon.
Sakanaka, Hidenori, 1989. *Kongo no Shutsunyukoku Kanri no Arikata ni tsuite*. Tokyo: Kodansha.
Sakata, Yasuyo, 2011. 'Korea and the Japan–US Alliance: A Japanese Perspective,' in Inoguchi Nakahashi, G. John Ikenberry and Yochiro Sato (eds), *The US-Japan Security Alliance, Regional Multilateralism*. New York: Palgrave Macmillan.
Sandusky, Michael C., 1983. *America's Parallel*. Alexandria: Old Dominion Press.
Schaller, Michael, 1997. *Altered States: The United States and Japan since the Occupation*. New York: Oxford University Press.
Schilling Mark, 2012. 'Kazoku no Kuni (Our Homeland),' *Japan Times*, 27 July 2012.
Segal, Gerald, 1991. 'Northeast Asia: Common Security or a la Carte?,' *International Affairs* 67(4): pp. 755–67.
Shain, Yossi, 2007. *Kinship and Diasporas in International Affairs*. Ann Abbor: The University of Michigan Press.
Shain, Yossi and Aharon Barth, 2003. 'Diasporas and International Relations Theory,' *International Organization* 57(3): pp. 449–79.
Sheffer, Gabriel, 2003. *Diaspora Politics: At Home Abroad*. New York: Cambridge University Press.
Shinoda, Tomohito, 2011. 'Costs and Benefits: US-Japan Alliance,' in Inoguchi Nakahashi, G. John Ikenberry and Yochiro Sato (eds), *The US-Japan Security Alliance, Regional Multilateralism*. New York: Palgrave Macmillan.
Shipper, Apichai W., 2008. *Fighting for Foreigners: Immigration and Its Impact on Japanese Democracy*. Ithaca: Cornell University Press.
—— 12/2009. 'Nationalism of and against Zainichi Koreans in Japan,' *Asian Politics & Policy* 2(1): pp. 55–75.
Shultz, George P., 1993. *Turmoil and Triumph: My Years as Secretary of State*. New York.
Skrbis, Zlatko, 2007. 'The Mobilized Croatian Diaspora: Its Role in Homeland Politics and War,' in Hazel Smith and Paul B. Stares (eds), *Diasporas in Conflict: Peace-Makers or Peace-Wreckers?* Tokyo and New York: United Nations University Press.
Smith, A.D., 1999. *Myths and Memories of the Nation*. Oxford and New York: Oxford University Press.
Smith, Hazel, 2007. 'Diasporas in International Conflict,' in Hazel Smith and Paul B. Stares (eds), *Diasporas in Conflict: Peace-Makers or Peace-Wreckers?*. Tokyo and New York: United Nations University Press.
Smith, Hazel and Paul B. Stares, 2007. *Diasporas in Conflict: Peace-Makers or Peace-Wreckers?*. Tokyo and New York: United Nations University Press.
Snyder, Glenn H., 1984. 'The Security Dilemma in Alliance Politics,' *World Politics* 36(4): pp. 461–96.

BIBLIOGRAPHY

—— 1990. 'Alliance Theory: A Neorealist First Cut,' *Journal of International Affairs* 44(1): pp. 103–23.
—— 1991. 'Alliances, Balance, and Stability,' *International Organization* 45(1): pp. 121–42.
Snyder, Scott, 2009. *China's Rise and the Two Koreas*. London: Lynne Rienner Publishers.
—— 2011. 'Korea and the US–Japan Alliance: An American Perspective,' in Inoguchi Takahashi, G. John Ikenberry and Yochiro Sato (eds), *The US-Japan Security Alliance, Regional Multilateralism*. New York: Palgrave Macmillan.
Sŏ, Sang-ch'ŏl, 1978. *Growth and Structural Changes in the Korean Economy, 1910–1940*. Cambridge, MA: Harvard University Press.
Soderberg, Marie, 2011. 'Japan-South Korea Relations at a Crossroads,' in Marie Soderberg (ed.), *Changing Power Relations in Northeast Asia*. London: Routledge.
Soon, Sung Cho, 1967. *Korea in World Politics, 1940–1950: An Evaluation of American Responsibility*. Berkeley: University of California Press.
Stratton, Samuel S., 1948. 'The Far Eastern Commission,' *International Organization* 2(1): pp. 1–18.
Stueck, William Whitney, 1981. *The Road to Confrontation: American Policy toward China and Korea, 1947–1950*. Chapel Hill: University of North Carolina Press.
Suh, dae-sook, 1967. *The Korean Communist Movement 1918–1948*. Princeton: Princeton University Press.
Suh, Sang-Chul, 1978. *Growth and Structural Changes in the Korean Economy 1910–1940*. London: Harvard University Press.
Suny, Ronald Grigor, 2001. 'Constructing Primordialism: Old Histories for New Nations,' *The Journal of Modern History* 73(4): pp. 862–96.
—— 2001. 'The Empire Strikes Out,' in Ronald Grigor Suny and Terry Martin (eds), *A State of Nations: Empire and Nation-Making in the Age of Lenin and Stalin*. Oxford and New York: Oxford University Press.
—— 2011. *The Soviet Experiment: Russia, the USSR, and the Successor States*, second edn. New York: Oxford University Press.
Tai, Eika, 2004. 'Koriakei Nihonjin to iu sentaku (A Choice to be Korean Japanese)', *Chuokoron*, pp. 360–2.
—— 2009. 'Between Assimilation and Transnationalism: The Debate on Nationality Acquisition among Koreans in Japan,' *Social Identities* 15(5): pp. 609–29.
Takahashi, Kosuke, 2014. 'Shinzo Abe's Nationalist Strategy With his overt Nationalism and his Historical Revisionism, Shinzo Abe has a Plan for Japan,' *The Diplomat*, 13 February 2014.
Takemae, E., R. Ricketts, S. Swann and J.W. Dower, 2002. *The Allied Occupation of Japan*. New York and London: Continuum.
Tamura, T., 1999. *Korean Population in Japan 1910–1945*, (1)-(3). Keizai-to Keizaigakusha.
—— 2003. 'The Status and Role of Ethnic Koreans in the Japanese Economy,' in C. Fred Bergsten and In-bŏm Ch'oe (eds), *The Korean Diaspora in the World Economy*. Washington, DC: Institute for International Economics.
Taylor, Gaubatz Kurt, 1996. 'Democratic States and Commitment in International Relations,' *International Organization* (1): pp. 109–39.
Terada, Takashi, 2011. 'Australia-Japan Security Partnership,' in Inoguchi Takahashi, G. John Ikenberry and Yochiro Sato (eds), *The US-Japan Security Alliance, Regional Multilateralism*. New York: Palgrave Macmillan.
Tilly, Charles, 1997. 'Means and Ends of Comparison in Macrosociology,' *Comparative Social Research* 16: pp. 43–53.
Tölölyan, Khachig, 1991. 'The Nation-State and Its Others: In Lieu of a Preface,' *Diaspora: A Journal of Transnational Studies* 1(1): pp. 3–7.

BIBLIOGRAPHY

────── 1991. 'Rethinking Diasporas: Stateless Power in the Transnational World,' *Diaspora: A Journal of Transnational Studies* 5(1): pp. 3–36.

────── 2007. 'The Armenian Diaspora and the Karabagh Conflict,' in Hazel Smith and Pail B. Stares (eds), *Diasporas in Conflict: Peace-Makers or Peace-Wreckers?* Tokyo and New York: United Nations University Press.

Tshiyuki, Tamura, 2003. 'The Status and Role of Ethnic Koreans in the Japanese Economy' in C. Fred Bergsten and Ch'oe In-bŏm (eds), *Korean Diaspora in the World Economy*. Washington, DC: Institute for International Economics.

Tsuboe, Senji, 1965. *Zai Nichi Chosenjin Undo No Gaikyo*. Tokyo: Homu kenkyu hokoku, 46–3.

Tsuda, Takeyuki, 2008. 'Local Citizenship and Foreign Workers in Japan,' *Japan Focus*.

Tsurushima, Setsure, 1978. 'Korean Immigrants in Kando in the 1920s,' *The Kansai University Review of Economics and Business* 7: pp. 48–52.

Uemura, Takashi, 1998. 'Open on Japanese Culture,' *Asahi Newspaper*, 20 October 1998, p. 2, http://database.asahi.com/library2e/main/start.php (accessed 14 March 2014).

Ulam, Adam B., 1968. *Expansion and Coexistence: The History of Soviet Foreign Policy, 1917–1967*. New York: Praeger.

Valencia, Mark, 2014. 'Asian Threats, Provocations Giving Rise to Whiffs of War,' *Japan Times*, 9 June 2014.

Van, den Berghe Pierre L., 1983. 'Class, Race and Ethnicity in Africa,' *Ethnic and Racial Studies* 6(2): pp. 221–36.

Wada, Haruki, 2014. *The Korean War: An International History*. New York: Roman & Littlefield.

Wagner, E.W., 1951. *Korean Minority in Japan 1904–50*. New York: Institute of Pacific Relations.

Walt, Stephen M., 1985. 'Alliance Formation and the Balance of Power,' *International Security* 9(4): pp. 3–43.

────── 1987. *The Origins of Alliances*. Ithaca: Cornell University Press.

────── 1997. 'Why Alliances Endure or Collapse,' *Survival* 39(1): pp. 156–79.

Waltz, Kenneth, 1979. *Theory of International Politics*. Massachusetts: Addison-Wesley.

────── 1986. 'Anarchic Orders and Balances of Power,' in Robert O. Keohane (ed.), *Neorealism and its Critics: The Political Economy of International Change*. New York: Columbia University Press.

Watanabe, Akio, 1985. 'Political Change in Japan and Korea-Japan Relations,' in *Korea and Japan in World Politics*. Seoul: The Korean Association of International Relations.

Watson, Robert J., 1986. *History of the Joint Chiefs of Staff*, vol. 5: *The JCS and National Policy, 1952–1954*. Washington, DC.

Watt, Lori, 2009. *When Empire Comes Home: Repatriation and Reintegration in Postwar Japan*. Cambridge: Harvard University Press.

Weber, Eugen. 1976. *Peasants into Frenchmen: The Modernization of Rural France, 1870–1914*. Stanford: Stanford University Press.

Weinberger, Caspar, 1986. 'US Defence Strategy,' *Foreign Affairs* 64(4): pp. 675–97.

Weiner, Michael, 1989. *The Origins of the Korean Community in Japan, 1910–1923*. Manchester: Manchester University Press.

────── 1997. 'The Invention of Identity: 'Self' and 'Other' in Pre-War Japan,' in Micheal Weiner (ed.), *Japan's Minorities: The Illusion of Homogeneity*. London and New York: Routledge.

Weiner, Myron, 1971. 'The Macedonian Syndrome: An Historical Model of International Relations and Political Development,' *World Politics* 23(4): pp. 665–83.

Welfield, John. 1988. *An Empire in Eclipse: Japan in the Postwar American Alliance System: A Study in the Interaction of Domestic Politics and Foreign Policy*. London and Atlantic Highlands, NJ: Athlone Press.

BIBLIOGRAPHY

Wendt, Alexander, 1994. 'Collective Identity Formation and International State,' *The American Political Science Review* 88(2): pp. 384–96.

The White House Office of the Press Secretary, 2014. 'Obama, Republic of Korea's President, Japanese PM in The Hague,' 25 March 2014, http://iipdigital.usembassy.gov/st/english/texttrans/2014/03/20140325296883.html?CP.rss=true#ixzz35Hgi1wsJhttp://iipdigital.usembassy.gov/st/english/texttrans/2014/03/20140325296883.html?CP.rss=true#axzz35HbvTloD (accessed 2 December 2015).

Whiting, Allen. S., 1968. *China Crosses the Yalu: The Decision to Enter the Korean War.* Stanford, CA: Stanford University Press.

Williamson, Lucy, 2014. 'Why is China's Leader visiting Seoul?,' BBC World News, 3 July 2014, http://www.bbc.co.uk/news/world-asia-28140036 (accessed 9 July 2014).

Wimmer, Andreas, 2002. *Nationalist Exclusion and Ethnic Conflict: Shadows of Modernity.* Cambridge: Cambridge University Press.

Yamaguchi, Noboru, 1999. 'Trilateral Security Cooperation: Opportunities, Challenges, and Tasks,' in Ralph A. Cossa (ed.), *US-Korea-Japan Relations: Building Toward a Virtual Alliance.* Washington, DC: The CSIS Press.

Yamazaki, Jane W., 2006. *Japanese Apologies for World War II.* London: Routledge.

Yoshida, Shigeru, 2007. *Yoshida Shigeru: Last Meiji Man.* Lanham: Rowman & Littlefield.

Index

1991 question 175, 176
38th parallel 19, 74, 75, 76, 83, 84, 85, 99

abandonment 38, 41, 47, 166, 167
accommodation 9, 13, 17, 23, 29, 32, 33, 43, 46, 116, 133, 134, 137, 143, 147, 157, 159, 162, 164, 177, 193, 205, 211, 212, 213, 215, 216, 217, 220, 223
Akihito 170
Alien Registration Law 34, 104
Alien Registration Ordinance 101
alliance cohesion 8, 9, 31, 36, 38, 39, 43, 45, 46, 48, 49, 116, 117, 125, 160–2, 164, 165, 166, 172, 176, 178, 181, 183, 193, 194–6, 205, 207, 211–13, 215, 216
annexation of Korea 53, 56, 184
Anti-Ballistic Missile (ABM) 128
Appadurai, Arjun 207
armistice 94, 173, 207
assimilation 2, 9, 15, 18, 26, 29, 32, 33, 43, 46, 57, 58, 61, 62, 64, 65, 68, 70, 71, 116, 156, 162, 164, 181, 205, 211, 212
asymmetry 116
Auslin, Michael 168

bilateralism 74
Bhabha, Homi 207
Brubaker, Rogers 7, 8, 15, 16, 26, 35, 39, 136, 156, 211
Brzenzinski, Zbigniew 142

Cairo Declaration 93, 99
Carter, Jimmy 139, 140–2, 149, 160
Cha, Victor 38, 45, 47, 114
Chen, Edward 57
Cheongjin 110
Chiang Kai shek 4, 86, 105
chima chogori 182
Chogin (Korean bank) 219
Chongryun (pro-North) 2, 3, 7, 11, 12, 13, 19, 20, 43, 107, 108, 111, 125, 134, 135, 137, 138, 143, 144, 145, 146, 159, 175–9, 182, 193, 194, 197–202, 212, 214–21
Chongryun remittances 178, 179
Chosen Sotoku 56
Chun Doo Hwan 149, 152
Churchill, Winston 74, 93
citizenship 103, 107, 123, 146, 156, 180, 181, 209, 213
collective self-defence 91, 191, 197, 204
colony 27, 56, 57, 58, 61, 67, 113
comfort women 169, 170
Cossa, Ralph 169
critical junctures 50
Cumings, Bruce 60, 67, 69, 70

dagger thrust 55, 91
Dairen 80
democratization 136, 146, 170
deportation 33, 103, 104, 105, 113, 125, 158, 213
Deng Xiaoping 142
détente 38, 127, 130, 132, 133, 138, 141, 149
Dokto, Takeshima 127
Dulles, John Foster 34, 74, 88, 91, 92, 94–6, 103, 112, 141

Eisenhower, Dwight D. 94, 95
entrapment 38, 167
Erin, Jenne 43
Etorofu 95
Exceptional Permanent Residence 155, 175, 214
exclusion 3, 9, 13, 16, 17, 18, 29, 32, 33, 43, 44, 46, 58, 78, 79, 92, 98, 101, 102, 105, 112, 113, 116, 117, 148, 162, 164, 178, 181, 182, 196, 200, 205, 207, 211–13, 216–18, 221, 222, 223
exclusive economic zone (EEZ) 169

INDEX

Far East 33, 83, 87, 91, 92, 93, 94, 95, 96, 97, 102, 104, 140
Far East Commission 104
Flying Horse Campaign 109
Fukuda, Takeo 143

Gaubatz, Kurt Taylor 168
Gellner, Ernest 24, 221
geopolitical blindness 160, 207
geopolitics 2, 6, 8, 9, 13, 14, 16, 18, 19, 25, 27, 28, 39, 40, 41, 43, 46–8, 50, 64, 76–9, 81, 84, 98, 99, 117, 125, 127, 160–5, 170, 176, 177, 181, 193, 195, 206, 207, 210, 211, 216
GHQ 99, 101
Gluck, Carol 160, 170, 207
Great Powers 79, 87, 88, 89, 98, 99, 207
Griffin, Christopher 168
Guam Doctrine 121

Habib, Philip 132
Habomais 95
Hagel, Chuck 190, 192, 205
Hallyu boom 179
Hammar, Thomas 123
Hatoyama, Ichiro 89, 94, 96, 105–9, 113
Hobsbawm, Eric 54
Host Nation Support (HNS) 139, 142

Ikeda, Hayato 112, 120
imperialism 63, 120, 124, 216
Inoue, Masutaro 107
International Committee of the Red Cross (ICRC) 108, 109, 110, 111
International Covenant on Human Rights 155
Ishihara, Shintaro 199
Ito, Hirobumi 55

Japan–US Security Treaty 88, 89, 91, 94–8, 109, 112, 113, 121, 139, 222
Japanese Communist Party (JCP) 42
Japanese Defence Agency (JDA) 153
Joint Commission 81
Joint US Military Assistance Group–Korea (JUSMAG-K) 143
Josŏnjok (Korean diaspora in China) 4
JSP (Social Democratic Party) 177
Juche (subject/self-reliance) 137

Kaifu, Toshiki 176
Kanemaru, Shin 142, 172, 177
Kang, Sang-jung 136, 146, 207, 208
Kaplan, Stephan 87
Khrushchev, Nikita 84
Kim Dae Jung 170, 174, 179
Kim Dong Jo 135
Kim Hyon Hui 172
Kim Il Sung 4, 19, 69, 84, 85, 87, 106, 107, 109, 132–4, 137, 143, 172, 197, 215, 216
Kim Jong Il 185, 186, 194, 197, 199, 217
Kim Jong Pil 140, 141
Kim Kyu Il 111, 112
Kim Young Sam 169, 73, 174, 179
Kirkwood, Montague 58
Kishi, Nobusuke 89, 96, 97, 98, 108–9, 112, 120
Kissinger, Henry 128, 129, 130
Klemens, Meckel 55
Kominka 65, 67
Korea Clause 121, 122, 126, 131, 139, 140, 160
Korea Energy Development Organization (KEDO) 179, 178, 173
Korean Central Intelligence Agency (KCIA) 123, 145
Korean Democratic Party 82
Korean People's Republic 82
Korean War 12, 42, 69, 70, 77, 84, 85, 86, 87, 88, 92, 100, 111, 128, 145, 164, 189, 207
Koryo Saram 4
Kunashiri 95
Kurils 93, 94, 95
Kwangju Uprising 152

League of Korean Residents (*Choren*) 5, 11, 12, 42, 102, 238
Liberal Democratic Party 96, 97, 108
Lie, John 1
Liska, George 73
livelihood protection 106

MacArthur, Douglas 88, 90, 101, 102, 103, 106, 112, 150
McCarran Walter Act 104, 222
Manchuria 25, 52, 53, 55, 56, 63, 64, 66, 69, 71, 76, 80, 83, 84, 93, 96, 98
Mao Ce-tung 87, 129
Mangyongbong (ferry) 194, 198, 199, 219, 220

INDEX

March First Uprising (*Samil Undong*) 62
Meiji 1, 7, 24, 25, 51, 55, 57, 61
Miki, Takeo 140, 141, 143
Mindan (pro-South) 2, 3, 11, 12, 125, 146, 180, 198, 214, 215
Minobe, Ryokichi 133, 199
Minsen (Democratic Front for the Unification of Koreans in Japan) 12, 42
Morris-Suzuki, Tessa 105
Morrow, James 37, 38, 116, 129, 166, 171, 173
Motono, Ichiro 56
Mun Se-kwang 134, 135, 136, 202
Murphy, Robert 92
Mylonas, Harris 8, 9, 17, 31, 32, 33, 36, 43, 46, 50, 98, 105, 113, 156, 160, 196, 197, 205, 206, 210, 211

Nakasone, Yasuhiro 151–5, 158, 159
National Association for the Rescue of Japanese kidnapped by North Korea (*Sukukai*) 201
nation-building 1, 3, 4, 5, 7, 8, 9, 11, 15, 17, 24, 25, 31, 32, 35, 36, 46, 70, 98, 113, 116, 160, 196, 205–7, 210, 211
National Defence Programme Guidelines 191
NATO (North Atlantic Treaty Organisation) 91, 129
naturalization 2, 135, 147, 156, 181, 182, 222
Niigata 110, 199
Nixon Doctrine 47, 132
Nixon, Richard 127, 128, 129, 130, 132, 140, 149, 220
'Nixon-shock' 130
Nodong 164, 172, 178, 194
non-interference 74, 91, 98, 132
Non-Proliferation Treaty (NPT) 172, 194
nordpolitik 47, 176
normalization 74, 128, 131, 172, 218
Northern Territories 95
nuclear weapons 97, 182, 185, 188, 194, 218, 220

Obama, Barack 8, 187
Obuchi, Keizou 170, 179
Official Development Assistance (ODA) 151
Okinawans 7, 64
Okonogi, Masao 70
Okuno, Seisuke 152

One-China Policy 131, 138, 195
One-Korea Policy 38, 48, 131
Organization for Economic Co-operation and Development (OECD) 203

Pachinko 177, 178, 248
Park Chung Hee 66, 98, 120, 123, 124, 145, 202
Patterson, Robert P. 83
permanent alien status 103
Peterson, Howard C. 83
proactive pacifism 204

Radford, Admiral Arthur W. 94
Reagan, Ronald 149, 150, 152
rearmament 77, 89, 91, 92, 94
Red Cross 105, 106, 107, 108, 110
re-entry visa 13, 134, 144
Reischauer, Edwin O. 97
repatriation 42, 71, 99, 103, 105–10, 112–14, 134, 210, 217, 218, 229
revisionism 32, 35, 106, 107, 113, 159
Rhee, Shingman 35, 68, 69, 82, 85, 88, 92, 93, 100, 103, 108, 214
Roh Moo-hyun 184
ROK–Japan Normalization Treaty 36, 47
Roosevelt, Franklin D. 68, 74, 80, 93
Ryukyu Island 92, 95, 120, 121, 126

Sakhalin Island 19, 20, 52, 53, 55, 66, 80, 93
Sakhalin Koreans 19, 20
San Francisco Treaty 95
Sato, Katsumi 201
SCAP (Supreme Commander of the Allied Powers) 99–103, 113
security and autonomy 31, 37, 38, 40, 43, 116, 118, 125, 132, 147, 213
Seikei, Bunri 120, 193, 195, 212
Self-Defence Force (SDF) 153, 191, 196, 204
self-determination 74
Senkaku Island 190, 191, 193, 197, 205
Seung-Young Kim 6
Shanghai communique 128
Shigemitsu, Mamoru 94, 95
Shikotan 95
Sino-Japanese War 63, 65
Six Party Talks 183, 186, 187, 188
Snyder, Glenn 39
South China Sea 190

INDEX

Stalin, Joseph 89, 93
status quo 32, 33, 37, 47, 63, 73, 87, 88, 98, 100, 105, 113, 115, 121, 190, 206
Strategic Arms Limitation Talks (SALT) 128
Strategic Defence Initiative (SDI) 149, 153
Stueck, William Whitney 80
Subversive Activities Prevention Law 104
Système de Rattachement 56

Taepodong ballistic missile 6, 174, 179
Taepodong launch 174, 179
Taiwan 8, 55–8, 65, 67, 87, 105, 121, 131, 195
Tanaka, Kakuei 8, 131, 135, 141, 143, 195
Terauchi, Masatake 56, 59, 62
third-country nationals (*daisan-kokujin*) 101
threat perception 127, 138, 159, 160, 161, 164, 165
Tojo, Hideki 96, 197
Treaty of Portsmouth 55
Treaty of Shimonoseki 55
Truman Doctrine 90
Truman, Harry S. 75, 80, 81, 85, 86, 88, 90, 91, 100
trusteeship 74, 80, 81, 98
Two Koreas approach 6, 13, 76, 112, 120, 158, 172, 196, 217, 220

United Nations Convention on the Law of the Sea (UNCLOS) 169
United Nations Temporary Commission on Korea 84
UN Refugee Convention 155
US Military Government in Korea 82
US–ROK Mutual Defence Treaty 78

Vienna Convention 134, 199
Vietnam War 150
virtual alliance 169, 182, 192

Weiner, Michael 51
Wilson, Woodrow 74
Walt, Stephen 36, 116, 127, 128, 129

Xi Jinping 203

Yalta 80, 93
Yenan 107, 109
Yoshida, Shigeru 34, 89, 90–4, 97, 100, 103–7, 113
Yushin 132

Zainichi identity 9, 10, 18, 146, 147, 180, 207, 208, 210
Zhou Enlai 131, 195